AF595132

Circular Economy in Low-Carbon Transition

Circular Economy in Low-Carbon Transition

Editors

Anna Mazzi
Jingzheng Ren

MDPI • Basel • Beijing • Wuhan • Barcelona • Belgrade • Manchester • Tokyo • Cluj • Tianjin

Editors
Anna Mazzi
Department of Industrial Engineering
University of Padova
PD - Padua
Italy

Jingzheng Ren
Department of Industrial and Systems Engineering
The Hong Kong Polytechnic University
Hong Kong
China

Editorial Office
MDPI
St. Alban-Anlage 66
4052 Basel, Switzerland

This is a reprint of articles from the Special Issue published online in the open access journal *Energies* (ISSN 1996-1073) (available at: www.mdpi.com/journal/energies/special_issues/Circular_Economy_Low-Carbon_Transition).

For citation purposes, cite each article independently as indicated on the article page online and as indicated below:

LastName, A.A.; LastName, B.B.; LastName, C.C. Article Title. *Journal Name* **Year**, *Volume Number*, Page Range.

ISBN 978-3-0365-3586-9 (Hbk)
ISBN 978-3-0365-3585-2 (PDF)

© 2022 by the authors. Articles in this book are Open Access and distributed under the Creative Commons Attribution (CC BY) license, which allows users to download, copy and build upon published articles, as long as the author and publisher are properly credited, which ensures maximum dissemination and a wider impact of our publications.
The book as a whole is distributed by MDPI under the terms and conditions of the Creative Commons license CC BY-NC-ND.

Contents

About the Editors

Anna Mazzi

Anna Mazzi is an Associate Professor of Environment Health & Safety Management Systems at the Department of Industrial Engineering, University of Padova, Padova, Italy. She is the head of the research group SAM.lab, dedicated to supporting organizations in sustainability assessment and management. She is a member of the Italian Network of Life Cycle Assessment and coordinates one of its working groups. During her research activity at University of Padova, she has collaborated in several European research projects, in the funding lines LIFE Environment, Interreg, Intelligent Energy Europe, EQUAL and CIP Innovation. She has also collaborated in numerous research projects at national and regional levels on the implementation of environmental assessment and management tools and standards at industrial and territorial levels. Her current research activity, documented by about 170 publications (h-index: 22), is focused on environmental impact evaluation and minimization, through life cycle sustainability assessments, to support process optimization and product innovation in reuse and recycling.

Jingzheng Ren

Dr. Jingzheng Ren is currently an Assistant Professor at The Hong Kong Polytechnic University. He is also an Honorary Associate Professor of Energy System Engineering at University of Southern Denmark (Denmark) and an Associated Senior Research Fellow at Institute for Security & Development Policy (Sweden). He serves as the Co-Editor-in-Chief of *Environment, Development and Sustainability* (Springer) and Editor of *Sustainable Production and Consumption* (Elsevier), and was previously an Associate Editor of *Journal of Renewable and Sustainable Energy* (AIP), Associate Editor of *Renewable Energy & Sustainable Development*, and a Guest Editor of several leading SCI indexed international journals with impact factors. He has more than 200 publications to his name, including 170 journal papers, 17 authored/edited books, and more than 50 book chapters. Eleven of his papers have been selected as the ESI top 1% highly cited papers and four have been highlighted as Key Scientific Articles contributing to excellence in energy research. One paper was recognized as one of the three top papers published in Sustainable Production and Consumption in 2018 and won the Best Paper Award (with a second award for contributions to a leading journal on sustainability).

Preface to "Circular Economy in Low-Carbon Transition"

Climate change, environmental pollution, the energy crisis and resource depletion have become more and more serious. The circular economy, as a new business model that is different from the economy, can achieve the reuse and recycling of waste for waste minimization, improve the efficiency of resource utilization, and mitigate carbon emissions. It is no doubt that promoting the development of the circular economy can facilitate the transition to low-carbon processes and carbon-neutral development. However, there are still several questions that need to be answered: (1) How can the circular economy contribute to a low-carbon transition? (2) How do we address the fact that the circular economy model may also cause some new environmental problems, and how should we identify what the most environmentally friendly solution is among multiple alternatives for the circular economy? (3) Governmental regulation, policies and incentives play a significant role in promoting the development of the circular economy, so what are the policy instruments that can contribute to its development? (4) How can technological progress and solutions contribute to the circular economy? (5) How can environmental impact assessments contribute to the circular economy? (6) How can we achieve a circular economy or low-carbon transition through changes in consumption behaviors? In order to answer the above-mentioned questions, we launched a Special Issue in *Energies*. There are a total of six papers published in this Special Issue. This e-book collects these papers to build a platform for sharing advanced concepts, tools and methods for the users to take actions to achieve a circular economy.

There are six chapters in this book. Chapter 1 is a short commentary about the circular economy and the book. Chapter 2 focuses on a comparative environmental assessment of heat pumps and gas boilers based on a life cycle tool, and it demonstrates how to identify the most environmentally friendly solution when there are multiple solutions for achieving a circular economy. Chapter 3 introduces a modern policy instrument, a digital product passport, to contribute to the development of a circular economy. Chapter 4 used Aspen Plus for the simulation of a biomass gasification system for combined heat and power and conducts an economic assessment of this system to show how technological solutions and tools can be used to achieve a circular economy in a low-carbon transition. Chapter 5 aims to assess the environmental impacts of short and long food supply chains in some EU countries and uses LCA to compare eco-efficiency. Chapter 6 presents a comprehensive bibliometric analysis of carbon labeling schemes from 2007 to 2019, and the readers can learn how carbon labeling schemes can help to change human behaviors for carbon emission mitigation.

Anna Mazzi and Jingzheng Ren
Editors

 MDPI

Editorial

Circular Economy in Low-Carbon Transition

Anna Mazzi [1,*] and Jingzheng Ren [2]

1 Department of Industrial Engineering, University of Padova, Via Marzolo 9, 35131 Padova, Italy
2 Department of Technology and Innovation, University of Southern Denmark, Campusvej 55, 5230 Odense, Denmark; jire@iti.sdu.dk
* Correspondence: anna.mazzi@unipd.it; Tel.: +39-049-827-1611

Citation: Mazzi, A.; Ren, J. Circular Economy in Low-Carbon Transition. *Energies* **2021**, *14*, 8061. https://doi.org/10.3390/en14238061

Received: 24 November 2021
Accepted: 30 November 2021
Published: 2 December 2021

Publisher's Note: MDPI stays neutral with regard to jurisdictional claims in published maps and institutional affiliations.

Copyright: © 2021 by the authors. Licensee MDPI, Basel, Switzerland. This article is an open access article distributed under the terms and conditions of the Creative Commons Attribution (CC BY) license (https://creativecommons.org/licenses/by/4.0/).

The circular economy represents a fundamental pillar for modern business models and sustainable development targets: the mandatory claim "reduce, reuse, recycle" is the answer to the global criticalities of natural resources depletion and waste increase [1,2]. At the same time, energy production and consumption play key roles in the face of challenges of industrialization and rapid population growth: with the depletion of traditional fossil fuels, renewable and low-carbon energy sources have attracted more and more attention for their advantages such as high renewability, great development potential, and possible emissions-mitigation [3,4]. To implement the circular economy in the low-carbon transition, new supply chain opportunities can be explored; at the same time, new dilemmas must be carefully solved through the life-cycle approach, to avoid the environmental burdens shifting [5,6]. The international community—including scientists, policymakers, industries, and markets—must develop new tools and competencies to support interdisciplinary innovation through the adoption of a comprehensive perspective, and to generate sustainable values from green low-carbon behavior [7,8].

This book contains the successful invited submissions [9–13] to the Special Issue of *Energies* (ISSN 1996-1073) on the subject area of "Circular Economy in Low-Carbon Transition" in the section "Energy Economics and Policy". This Special Issue contributes to outline a roadmap of circular economy in the low-carbon transition, through the exchange of experiences in different contexts with both environmental and socio-economic points of view.

We sincerely thank the editorial staff and reviewers for their efforts and help to collect, select, and review the papers. We believe that the published articles will inspire both scientists and practitioners to explore new directions to the circular economy in new carbon transition.

Qualitative and quantitative measurements in resources/energy utilization, multi-criteria impact assessment in energy systems, and closing the loop initiatives enrich the international debate relating the topic. New research trends underlined by this Special Issue encourage continued discussion about the role of energy policies and technologies to achieve the SDGs and the climate actions using a life-cycle approach. The common objective must be the overall reduction in impacts and the formulation of substantially sustainable solutions, rather than downloading the problems along the supply chain or postponing the damages in the next decades.

Author Contributions: Conceptualization, A.M. and J.R.; writing—original draft preparation, A.M.; writing—review and editing, J.R. All authors have read and agreed to the published version of the manuscript.

Funding: This research received no external funding.

Institutional Review Board Statement: Not applicable.

Informed Consent Statement: Not applicable.

Data Availability Statement: Not applicable.

Conflicts of Interest: The authors declare no conflict of interest.

References

1. Olivetti, E.A.; Cullen, J.M. Toward a sustainable materials system. *Science* **2018**, *360*, 1396–1398. [CrossRef] [PubMed]
2. European Commission. *Closing the Loop—An EU Action Plan for the Circular Economy. The Circular Economy Package Proposal;* European Commission: Brussels, Belgium, 2015; Available online: https://eur-lex.europa.eu/legal-content/EN/TXT/?uri=CELEX:52015DC0614 (accessed on 1 December 2021).
3. Davis, S.J.; Lewis, N.S.; Shaner, M.; Aggarwal, S.; Arent, D.; Azevedo, I.L.; Benson, S.M.; Bradley, T.; Brouwer, J.; Chiang, Y.-M.; et al. Net-zero emissions energy systems. *Science* **2018**, *360*, 9793. [CrossRef] [PubMed]
4. Chen, W.-H.; Ong, H.C.; Ho, S.-H.; Show, P.L. Green Energy Technology. *Energies* **2021**, *14*, 6842. [CrossRef]
5. Hellweg, S.; Canals, L.M. Emerging approaches, challenges and opportunities in life cycle assessment. *Science* **2014**, *344*, 1109–1113. [CrossRef] [PubMed]
6. Sovacool, B.K.; Ali, S.H.; Bazilian, M.; Radley, B.; Nemery, B.; Okatz, J.; Mulvaney, D. Sustainable minerals and metals for a low-carbon future. *Science* **2020**, *367*, 30–33. [CrossRef] [PubMed]
7. Gonçalves, A.; Silva, C. Looking for Sustainability Scoring in Apparel: A Review on Environmental Footprint, Social Impacts and Transparency. *Energies* **2021**, *14*, 3032. [CrossRef]
8. Zhao, R.; Wu, D.; Patti, S. A Bibliometric Analysis of Carbon Labeling Schemes in the Period 2007–2019. *Energies* **2020**, *13*, 4233. [CrossRef]
9. Xiao, J.; Zhen, Z.; Tian, L.; Su, B.; Chen, H.; Zhu, A.X. Green behavior towards low-carbon society: Theory, measurement and action. *J. Clean. Prod.* **2021**, *278*, 123765. [CrossRef]
10. Majewski, E.; Komerska, A.; Kwiatkowski, J.; Malak-Rawlikowska, A.; Wąs, A.; Sulewski, P.; Gołaś, M.; Pogodzińska, K.; Lecoeur, J.-L.; Tocco, B.; et al. Are Short Food Supply Chains More Environmentally Sustainable than Long Chains? A Life Cycle Assessment (LCA) of the Eco-Efficiency of Food Chains in Selected EU Countries. *Energies* **2020**, *13*, 4853. [CrossRef]
11. Almpantis, D.; Zabaniotou, A. Technological Solutions and Tools for Circular Bioeconomy in Low-Carbon Transition: Simulation Modeling of Rice Husks Gasification for CHP by Aspen PLUS V9 and Feasibility Study by Aspen Process Economic Analyzer. *Energies* **2021**, *14*, 2006. [CrossRef]
12. Adisorn, T.; Tholen, L.; Götz, T. Towards a Digital Product Passport Fit for Contributing to a Circular Economy. *Energies* **2021**, *14*, 2289. [CrossRef]
13. Sevindik, S.; Spataru, C.; Aparisi, T.D.; Bleischwitz, R. A Comparative Environmental Assessment of Heat Pumps and Gas Boilers towards a Circular Economy in the UK. *Energies* **2021**, *14*, 3027. [CrossRef]

MDPI

Article

A Comparative Environmental Assessment of Heat Pumps and Gas Boilers towards a Circular Economy in the UK †

Selman Sevindik [1,*], Catalina Spataru [1], Teresa Domenech Aparisi [2] and Raimund Bleischwitz [2]

1 Energy Institute, Bartlett School Environment, Energy and Resources, University College London, London WC1E 6BT, UK; c.spataru@ucl.ac.uk
2 Institute for Sustainable Resources, Bartlett School Environment, Energy and Resources, University College London, London WC1E 6BT, UK; t.domenech@ucl.ac.uk (T.D.A.); r.bleischwitz@ucl.ac.uk (R.B.)
* Correspondence: s.sevindik@ucl.ac.uk

† This paper is an extended version of our conference papers from 15th SDEWES Conference, Cologne, Germany, 1–5 September 2020 and ICEESEN Conference, Kayseri, Turkey, 19–21 November 2020.

Citation: Sevindik, S.; Spataru, C.; Domenech Aparisi, T.; Bleischwitz, R. A Comparative Environmental Assessment of Heat Pumps and Gas Boilers towards a Circular Economy in the UK. *Energies* **2021**, *14*, 3027. https://doi.org/10.3390/en14113027

Academic Editor: Anna Mazzi

Received: 1 April 2021
Accepted: 14 May 2021
Published: 24 May 2021

Publisher's Note: MDPI stays neutral with regard to jurisdictional claims in published maps and institutional affiliations.

Copyright: © 2021 by the authors. Licensee MDPI, Basel, Switzerland. This article is an open access article distributed under the terms and conditions of the Creative Commons Attribution (CC BY) license (https://creativecommons.org/licenses/by/4.0/).

Abstract: This research compares the potential environmental impacts of heat pumps with gas boilers and scenario analysis through utilising the life cycle approach. The study analyses the current situation with the baseline model and assesses future applications with Circular Economy (CE), Resource Efficiency (RE) and Limited Growth (LG) scenarios. Then, hybrid applications of low-carbon technologies and different manufacturing scenarios are investigated according to baseline and CE scenarios. Our results show that the use and manufacturing phases are responsible for 74% and 14% of all environmental impacts on average as expected. Even though the electricity mix of the UK has decarbonised substantially during the last decade, heat pumps still have higher lifetime impacts than gas boilers in all environmental categories except climate change impact. The carbon intensity of heat pumps is much lower than gas boilers with 0.111 and 0.097 kg CO_2e for air source heat pumps and ground source heat pumps, whereas the boiler stands as 0.241 kg CO_2e. Future scenarios offer significant reductions in most of the impact categories. The CE scenario has the highest potential with a 44% reduction for heat pumps and 27% for gas boilers on average. RE and LG scenarios have smaller potential than the CE scenario, relatively. However, several categories expect an increase in future scenarios such as freshwater ecotoxicity, marine ecotoxicity and metal depletion categories. High deployment of offshore wind farms will have a negative impact on these categories; therefore, a comprehensive approach through a market introduction programme should be provided at the beginning before shifting from one technology to another. The 50% Hybrid scenario results expect a reduction of 24% and 20% on average for ASHP and GSHP, respectively, in the baseline model. The reduction is much lower in the CE scenario, with only a 2% decrease for both heat pumps because of the reduction in heat demand in the future. These results emphasise that even though the importance of the use phase is significant in the baseline model, the remaining phases will play an important role to achieve Net-Zero targets in the future.

Keywords: built environment; circular economy; gas boilers; heat pumps; life cycle assessment

1. Introduction

In 2018, 55% of the global population lived in cities, and this is expected to reach 68% by 2050 [1]. Cities are globally responsible for 75% of primary energy consumption [2] and 60–80% of greenhouse gas emissions [3]. Of all industrial sectors, the built environment is responsible for 36% of global energy consumption and 39% of energy-related greenhouse gas (GHG) emissions [4]. Use emissions such as heating, cooling, lighting, and cooking account for 72% of these emissions and the remaining comes from embodied emissions. Building-related emissions have decreased by 13% since 2013 and are around 20% below 1990 levels in the UK [5].

Heating is responsible for nearly half of UK energy usage and a third of carbon emissions, and currently 70% of heating purposes provided by natural gas [6]; therefore, electrification of the heating scheme is crucial via low-carbon technologies. In 2017, The Clean Growth Strategy was introduced in the UK, which was preceded by The Climate Change Act in 2008. The main proposals for the strategy aim to reduce UK emissions through energy efficiency in industry and housing, low-carbon transportation, clean power generation and enhancing natural resources, which account for 38%, 24%, 21% and 15% of UK emissions, respectively [6].

Efficiency improvements in buildings have been the major strategy during the last half-century. Very low heat conductivity in the building fabric is provided with low U-value insulation materials through passive strategies. Thermal mass also became a common topic in order to reduce heating and cooling loads and shifting peak demand. It is possible to limit the indoor temperature fluctuation to ± 4 °C with a heavyweight construction which helps to reduce discomfort in buildings [7]. On the other hand, ambitious building regulations have created a demand for low-exergy heating systems such as heat pumps. It is possible to reach energy and GHG emissions saving with a low-temperature hydronic heating system, even though it is a gas boiler. However, heat pumps maximise the saving as their efficiencies are higher in low temperatures [8].

A heat pump is a low-carbon technology that exploits heat from air, ground or water sources by heat transfer and provides heating, cooling, and domestic hot water (DHW). It could utilise electricity, mechanical or thermal energy in various applications such as residential, commercial, industrial or district heating. An electricity-driven heat pump could provide a three to four times higher amount of heat than the electricity consumed; therefore, it is expected to play a significant role in the decarbonisation of heating in buildings [9].

Heat pumps offer higher efficiencies than gas boilers; however, various refrigerants perform differently in various evaporator and condenser temperatures; therefore, choosing the right refrigerant according to system description and temperature requirements is crucial [10]. Current heat pumps in the UK market utilise R410A refrigerant, which has high global warming potential (GWP 2088); however, the use of R134a (GWP 1300) and R32 (GWP675) has also been increasing [11]. The number of studies investigating natural refrigerants such as ammonia (GWP 0.1) has increased, and the results show that ammonia could be used as a refrigerant in both large and small applications; however, the cost of the system is more expensive than traditional ones [12]. Therefore, more support is needed from the government in order to introduce the system to the market.

Energy efficiency through low-carbon heating is one of the key policies requiring the improvement of the standards of 1.2 million new boilers installed each year in England, including the installations of control devices to save energy. Moreover, reforming the Renewable Heat Incentive (RHI) and spending GBP 4.5 billion to support low-carbon heating technologies is expected between 2016 and 2021. GBP 184 million of investment has been scheduled for innovations in energy efficiency and low-carbon heating options [6]. However, the UK Government was planning to replace RHI according to the new Net-Zero target for 2050; therefore, RHI is extended until March 2022 and consultation for a new support scheme has been introduced [13]. It aims to support energy efficiency and low-carbon heating in buildings with a GBP 9bn investment during the next ten years.

The total share of renewables in heating and cooling has been increasing during the last decade in the UK; however, it was still only 7.5% in 2017 and the UK was at the end of the list among the EU member states with the Netherlands (5.9%) and Ireland (6.9%) [14]. The total number of heat pumps reached 9.5 million in the EU, which represented 4% of the building stock and the capacity of 82.7 GW [15]. The highest number of heat pumps sold in 2017 in the EU was for France (240,000 units), while only 20,000 units were sold in the UK. According to the Climate Change Committee [16], this number should have been 30,000 in 2020 and much more ambitious long-term targets should be set by the government. The number of heat pumps sold in the UK is still low and the UK Government have plans for not only single heat pump applications but also hybrid use with gas boilers [17]. The CCC also suggested that new homes should not be connected to the gas grid after 2025 and hybrid applications of heat pumps should start and reach 10 million by 2035 [18].

Domestic RHI was introduced in 2014 and, according to Ofgem, in 5 years, 55,000 domestic heat pumps have been deployed [19]. According to the CCC (CCC, 2018), at least 2.5 million heat pumps need to be deployed by 2030 in order to continue further progress of decarbonisation [20]. New residential applications are the majority of the heat pump market, which was around 10,500 units per year. However, this represents a small proportion when it is compared with the number of new residential units completed, which is around 175,000 per year approximately [15]. Even though current numbers are quite limited for now, more incentives and advances in the manufacturing process could help to increase the deployment rate. The UK Government's ten-point Industrial Revolution plan aims to have 600,000 heat pump installations per year by 2028 [21]. Moreover, the UK Government started a consultation in 2019 for future home standards to upgrade Part L and Part F of the building regulations [22]. This consultation was concluded in January 2021 and proposed a timetable for the implementation of future home standards. A total of 70% of respondents to this consultation believe that heat pumps will play a significant role in this standard, and there is already support from stakeholders [23]. According to the UK Government, future buildings should have 75–80% fewer CO_2 emissions than current built ones with these standards [24]. The RIBA Council introduced a challenge for designers, architects and industry to reduce operational energy demand, embodied carbon and water use through higher benchmarks for buildings [25]. As space heating plays a significant role in operational energy and carbon emissions, the importance of heat pumps as a low-carbon technology becomes more crucial to reach these benchmarks. According to a study conducted by the Department of Energy and Climate Change, in a mass-market scenario, cost reductions of around 18% are expected compared to current costs for Ground Source Heat Pumps (GSHP) and 20% for Air Source Heat Pumps (ASHP) [26,27], which could lead to a higher deployment rate in the future. However, reuse and recycle options of these systems should be considered before moving to mass deployment. Therefore, there is a need for harmonisation of current built environment theory with the theory underpinning CE in order to achieve circular chains.

This study extends the analysis presented in two conferences [28,29]. The aim of this research is a comparison of different environmental impact categories for key technologies to decarbonise heating in domestic buildings in the UK. Heat pumps and gas boilers are key technologies in the decarbonisation of buildings and have been selected as a relevant case to test our hypotheses and methods. Their impacts on low-carbon heating targets have been assessed through a Life Cycle Assessment (LCA) analysis for the current year, and future scenarios have been developed to assess their environmental impacts through LCA to understand the impacts of the replacement of existing technologies with new ones. The functional unit of the study is decided as 'generating 1 kWh of thermal energy for domestic heating', but cumulative results have also been presented to investigate lifetime environmental burdens associated with these heating technologies. Hybrid applications of heat pumps with gas boilers also assessed as hybrid technologies will play a significant role in the future according to government targets.

2. Methods

Life Cycle Assessment (LCA) is an analytical tool to assess the environmental impacts of a product or process through analysing the entire life cycle (raw material acquisition, production, use and disposal phases). Its aim was to reduce cost while improving performance; therefore, it has been widely used during the last couple of decades [30]. A Life Cycle Assessment (LCA) approach has been undertaken to evaluate the environmental impacts of low-carbon heating technologies for the domestic sector in accordance with ISO 14,040 and ISO 14,044 standards [31,32]. The analysis has four stages: (i) defining goal and scope to identify the purpose of the study and products; (ii) inventory analysis to collect data of the unit processes of products and analyse; (iii) impact assessment to identify environmental impacts; and (iv) interpretation to evaluate results and compare with potential solutions. The first step of this study focuses on the current situation of heating technologies, then, future scenarios try to evaluate their impact according to government plans and targets.

2.1. *Goal and Scope*

The goal of this study is to evaluate the potential environmental impacts of residential space heating in the UK through developing life cycle models for an air-source heat pump (ASHP), ground-source heat pump (GSHP) and natural gas boilers (NGB). This comprises a scenario analysis with the objective of achieving the Net-Zero target by 2050.

The functional unit of the study is decided as 'generating 1 kWh of thermal energy for domestic heating'. However, cumulative results have also been presented to investigate lifetime environmental burdens associated with these heating technologies. The LCA software SimaPro 8.0.3 [33] has been used to model the products and the ReCiPe Midpoint (H) method [34] has been used to calculate environmental loads.

2.2. *Inventory Data and Assumptions*

2.2.1. System Description and Boundary

System specification and material requirements of heat pumps and gas boilers and data for these products have been taken from a previous study [35] that analysed the environmental implications of these products in the UK. Heat pumps are decided as air to water and ground to water, and heating is provided by underfloor heating. The underfloor heating system is designed as a screed system covering 150 m^2 area. Material requirements of heat pumps and gas boilers are illustrated (Table 1). The capacity of the systems and operation period have been assumed as 10 kW and 2000 h/year. The total space heating demand was assumed to be 20,000 kWh/year for both heat pumps and gas boilers, which represents an average UK household yearly heating demand. All technologies are considered maintenance free; however, it is assumed that the refrigerant needs to be topped up 6% yearly as losses occur. The total lifetime of both heat pumps and gas boilers was assumed as 20 years.

Table 1. Material requirements for heating technologies. Data source: [35].

		ASHP			GSHP				NGB
Material	**Unit**	**Heat Pump**	**Under-Floor Heating System**	**Maintenance**	**Heat Pump**	**Under-Floor Heating System**	**Heat Collector System**	**Maintenance**	**Gas Boiler**
Polyolester oil	kg	2.7			1.7				
R-134A	kg	4.9		5.9	3.1			3.7	
Rockwool	kg								8.0
Low-alloyed steel	kg	32.0			20.0				115.0
Reinforcing steel	kg	120.0			75.0		33.0		
Stainless Steel	kg	5.0							5.0
Bentonite	kg				3.8				
Sand	kg		4650.0			4650.0			
Cement	kg		900.0			900.0			
Copper	kg	36.6			22.0				3.0
Aluminium	kg		126.0			126.0			7.5
Brass	kg						6.6		0.1
Polyvinylchloride	kg	1.6			1.0				
HDPE	kg	0.5					301.2		0.9
LDPE	kg		101.0			101.0	4.7		
Polystyrene	kg		66.0			66.0			
Elastomere	kg	16.0			10.0				
Ethylene Glycol	kg						167.0		
Total	kg	219.3	5843.0	5.9	136.6	5843.0	512.5	3.7	139.5

The system boundary of gas boilers and heat pumps includes extraction and production of raw materials, transportation of raw materials for assembly, manufacturing of heat pumps and gas boilers, manufacturing of underfloor heating system for heat pumps, manufacturing of heat collector for GSHP, transportation of products to the distributor, transportation of products to the installation site, installation of GSHP as it requires drilling, operation period which includes natural gas processing for boilers and electricity generation for heat pumps, maintenance of refrigerant for heat pumps and disposal of materials (reuse, recycling, landfilling, etc.) (Table 2). The installation phase is only considered for GSHP as it requires drilling, which is an extensive installation when compared with ASHP and gas boilers. As two types of heat collector exist for GSHP (horizontal and vertical), this study only included the horizontal one for simplicity. The difference between the two types is the amount of pipework for heat collectors, the heat carrier liquid and the type of machines to dig the ground. The maintenance stage is only considered for heat pumps as there will be losses in refrigerant during the operation period; therefore, annual top-up is required. Additionally, the underfloor heating system is only included for heat pumps as replacing the gas boiler will require either resizing the radiators or the installation of an underfloor heating system to achieve higher efficiency. Therefore, in this study, the underfloor heating system is included in the system boundary of heat pumps. However, renewing the gas boiler does not require any system change; therefore, no new heating system is proposed.

Table 2. Processes in system boundaries included for heat pumps and gas boilers.

Processes in System Boundaries	ASHP	GSHP	NGB
Extraction and production of raw materials	•	•	•
Transportation of raw materials for assembly	•	•	•
Manufacturing of heating technologies	•	•	•
Manufacturing of underfloor heating system	•	•	
Manufacturing of heat collector		•	
Transportation of products to the distributor	•	•	•
Transportation of products to the installation site	•	•	•
Installation of the products		•	
Operation period including natural gas processing and electricity generation	•	•	•
Maintenance	•	•	
Disposal of products	•	•	•

2.2.2. Transport

Heat pump installations in the UK market heavily rely on imports. A total of 69% of ASHP and 59% of GSHP are manufactured outside of the UK [13]. Europe is the dominant market as 70% of imported products are manufactured there. When individual countries are investigated, Sweden has the highest imported heat pump amount followed by South Korea, Spain, Italy, Czech Republic, and Germany. This study, therefore, selects Europe as the manufacturing location for heat pumps. Ecoinvent generic values (100–200 km) have been used for raw materials and assembly transport assumptions [36]. Heat pumps are assumed to be manufactured in Europe and transported to the UK (Table 3). Within this process, raw materials are transported 200 km by railway and 100 km by a large truck (16–32 tonne). After the assembly of the heat pump, it is transported to the distributor 500 km by railway and 200 km by a large truck (16–32 tonne). Then, the installation site distance has been assumed as 200 km and the products have been transported by a small truck (3.5–7.5 tonne). The underfloor heating system (UHS) and heat collectors (HC) are assumed to be manufactured in the UK; therefore, transport distances for manufacturing have been assumed as 200 km by railway and 100 km by a large truck (16–32 tonne), and installation distance has been assumed as 200 km. Natural gas boilers are assumed to be manufactured in the UK; therefore, transport for raw materials has been assumed as 200 km by railway and 100 km by a large truck (16–32 tonne). Distances from manufacturer to distributor and installation site have been assumed as 200 km.

Table 3. Transport assumptions. Data source: [36].

	Transport of;		Rail	Truck (Large)	Truck (Medium)	Truck (Small)
Heat Pumps	raw materials to manufacturer	km	200	100		
	products from manufacturer to distributor	km	500	200		
	products for installation	km				200
Gas Boiler	raw materials to manufacturer	km	200	100		
	products from manufacturer to distributor	km			200	
	products for installation	km				200
Underfloor Heating System	raw materials to manufacturer	km	200	100		
	products for installation	km			200	
Heat Collectors	raw materials to manufacturer	km	200	100		
	products for installation	km		200		

2.3. Scenario Analysis

The study offers a scenario analysis to assess the environmental impacts of heat pumps and gas boilers through LCA to understand the implications of the replacement of existing technologies with new ones. In this section, three scenarios have been developed for the year 2050. In the next section (Section 2.4), three more alternative scenarios have been developed for hybrid applications of technologies and another transport scenario. The latter is separated from the first three because they are assessed according to both the baseline model and also the Circular Economy (CE) scenario.

Circular Economy (CE) scenario: High technology development and high consumer engagement are supported by policies; therefore, more efficient houses and low-carbon technologies expect a reduction in energy demand. The decarbonisation of electricity is provided by increased offshore wind capacity, and the share of natural gas is nearly eliminated. Larger roles for heat pumps are provided and gas boilers are replaced with low-carbon technologies such as stand-alone heat pumps (10.7 million) or hybrid heat pumps (7.1 million). The number of gas boilers will reduce to 5 million and the remaining heating demand will be provided by district heating and biomass. A reduction in material demand and better waste treatment options are assumed with high policy support (Table 4).

Table 4. Summary of system specifications and assumptions for scenarios.

Drivers	Baseline		Circular Economy (CE)		Resource Efficiency (RE)		Limited Growth (LG)		Sources
Recycling rates for materials	Steel	75%	Steel	100%	Steel	90%	Steel	80%	[35,37,38]
	Copper	61%	Copper	100%	Copper	85%	Copper	75%	
	Aluminium	69%	Aluminium	100%	Aluminium	90%	Aluminium	80%	
	Plastics	32%	Plastics	100%	Plastics	80%	Plastics	60%	
	Refrigerant	80%	Refrigerant	100%	Refrigerant	90%	Refrigerant	80%	
SPF and Efficiency	ASHP	2.8	ASHP	4.2	ASHP	4.0	ASHP	3.6	[17,39,40]
	GSHP	3.4	GSHP	4.6	GSHP	4.4	GSHP	4.2	
	NGB	90%	NGB	95%	NGB	95%	NGB	95%	
Efficiency improvements			25%		15%		8%		[19,20,27]
Heat pump deployment (million)	ASHP	0.126	ASHP	10.479	ASHP	5.731	ASHP	0.809	[19]
	GSHP	0.015	GSHP	0.178	GSHP	0.132	GSHP	0.089	
	Hybrid HP	0.016	Hybrid HP	7.065	Hybrid HP	2.705	Hybrid HP	0.833	
	Gas Boiler	21.989	Gas Boiler	5.196	Gas Boiler	5.861	Gas Boiler	22.138	

Resource Efficiency (RE) scenario: A reduction in energy demand is expected but this decrease is lower than the CE scenario. The decarbonisation level of electricity is similar to the CE scenario. The deployment of heat pumps is limited (8.5 million), and the number of gas boilers is similar to the CE scenario; therefore, applications of hydrogen play a significant role in this scenario. High technology development and policy support are expected, but consumer support is relatively limited compared to the CE scenario.

Limited Growth (LG) scenario: Limited energy efficiency and technology development is assumed; therefore, residential heat demand is expected to reduce with the lowest number among other scenarios. The decarbonisation of electricity is not finished, and the deployment of heat pumps is very limited; therefore, the majority of heating demand will be provided by gas boilers. Slow adaptation to circular economy principles and low consumer engagement are expected.

2.3.1. Electricity Mix

The use phase of heating technologies has a significant impact on LCA analysis, accounting for 74% of overall impacts; therefore, updating the electricity mix of the UK to the current situation would help to see more accurate results. The current electricity mix of the UK for the year 2018 has been identified for the use phase of heat pumps (Figure 1) [41]. In 2018, 40.2% of electricity was produced from natural gas. Nuclear and wind accounted for 19.9% and 17.4%, respectively. The remaining comes from bioenergy, solar and coal.

According to National Grid, more than one-third of UK electricity was produced from natural gas and offshore wind capacity, around 10 GW, in 2018 [19]; however, more deployment of wind energy is expected in the future. The UK Government has revised its offshore wind capacity target from 30 GW to 40 GW by 2030 [42]. Therefore, National Grid's electricity mix scenarios have been adopted for the UK's future electricity mix ('Community Renewables', 'Two Degrees', and 'Steady Progression' scenarios adapted to 'Circular Economy', 'Resource Efficiency' and 'Limited Growth' scenarios, respectively) (Figure 1). All three scenarios assume a significant increase in wind energy but in different shares. In 2050, the RE scenario assumes that 56% of electricity will be produced from wind energy and the remaining will come from nuclear, solar and bioenergy, which account for 19%, 8% and 7%, respectively. In the CE scenario, wind energy reaches 60% of total electricity production. Solar and bioenergy increase to 10% each; therefore, the share of nuclear energy reduces to 12%. The LG scenario, however, assumes the least wind energy share with 53%; therefore, natural gas still has a share of 10% of total electricity production.

2.3.2. Efficiency of Technologies

One of the main impacts on energy demand in heat pumps is the Coefficient of Performance (COP), which identifies the ratio of energy needed according to its efficiency. Seasonal Performance Factors (SPFs) represent the average COP for heat pumps during the heating season. According to the Department for Business, Energy and Industrial Strategy (BEIS) monthly reports since 2014, average ASHP and GSHP efficiencies are calculated as 3.2 SPF and 3.5 SPF [39]. These values vary for legacy applications and new installations. Legacy ASHP applications have an average of 2.5 SPF and new installations have 3.4 SPF. Legacy GSHP installations, on the other hand, have an average of 2.9 SPF and the new ones have 3.8 SPF. Field trials in the UK and Europe show similar results with an average SPF of 2.6 and 3.2 for ASHP and GSHP, respectively [17]. Therefore, average SPFs for heat pumps have been considered as 2.8 for ASHP and 3.4 for GSHP. The efficiency of NGB is considered as 90% for the baseline model (Table 4).

Current heat pump efficiencies vary between manufacturer test data and field trials. Correct sizing and better installation of heat pumps provide higher efficiencies; thus, it could be possible to reach manufacturers' test efficiencies in the future. Over the years, the efficiency of heat pumps is expected to increase with the help of advances in the market. The CCC assumes a 0.5 increase in the COP of heat pumps between 2020 and 2030 [40]. Therefore, future scenarios in this study assume higher efficiencies varying between 3.6–4.2 for ASHP and 4.2–4.6 for GSHP (Table 4). GSHPs are expected to have a lower increase in COP than ASHPs due to their high outlet temperature and modest difference in the ground temperature around the heat collector [43]. Therefore, efficiency improvement in GSHP is expected to be lower than in ASHP.

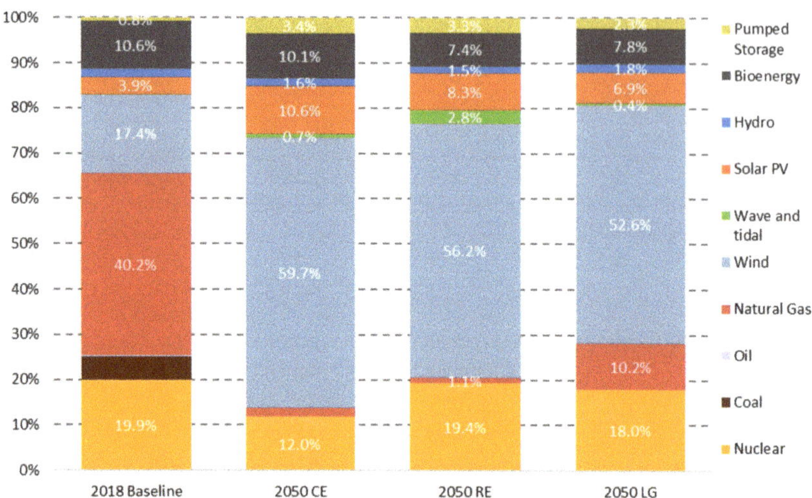

Figure 1. Electricity mix of UK in different scenarios used in LCA analysis. Data sources: [19,41].

2.3.3. Decommissioning

The lifetime of both technologies has been assumed to be 20 years, and at the end of their life cycle metal components are recycled and the rest is landfilled. UK and Europe recycling rates have been reviewed, and steel, copper, aluminium and plastics have been assumed as 75%, 61%, 69% and 32% recycled [37]. A total of 80% of refrigerants are assumed to be reused after 20% losses during the decommissioning [35].

Gas boilers and heat pumps are electrical and electronic equipment (EEE) covered by the WEEE Regulations under category 1 and category 12. They both have similar targets as 85% recovery and 80% recycling rates [38]. However, these benchmarks will likely increase if the UK continues to progress towards Net-Zero targets. Therefore, all scenarios expect an increase in the recycling rate; however, the CE scenario assumes the highest recycling rates with 100% for all components. High policy support and public engagement could help to achieve 100% recycle and recovery options.

2.3.4. Efficiency Improvements in Residential Sector

The need for space heating could be less in the future. Thermal efficiency improvements through retrofitting existing houses and setting higher benchmarks for new buildings mean that by 2050, domestic buildings could use up to 26 per cent less energy for heat compared to today [19]. The CCC [20] assumes a 15% reduction in energy consumption in the residential sector by 2030 through energy efficiency improvements in existing buildings. The Royal Institute of British Architects [27] has set a challenge for designers to reach at least a 75% reduction in operational energy in domestic buildings by 2030. Therefore, different measures have been taken for future scenarios. RE and CE scenarios assume a 15% and 25% reduction in heat demand in an average household. The LG scenario, however, does not consider any energy improvement measures as the economy faces limited economic growth (Table 4).

2.4. Hybrid and Transport Scenarios

In the previous section, model simulations are conducted based on individual heating technologies without focusing on hybrid applications. However, the UK Government and National Grid have decarbonisation targets for heating, and scenarios show that there will be a need for hybrid options in the future. Additionally, Asia is a dominant market, and some companies manufacture their heat pumps in Asian countries. Moreover, South Korea is the second country that the UK has the highest heat pump imports from [13]. Therefore, this section investigates the impact of changing the manufacturing location and hybrid options according to the baseline scenario and CE scenario, which was modelled in the previous section. Three scenarios are investigated as:

- Transport (SK) scenario assumes that heat pumps are manufactured in Asia and average ROW (rest of the world) production values have been used in SimaPro. South Korea has been chosen as a manufacturing country to identify the main shipment method and distance as a transoceanic freight shipment of 12,400 nm (22,965 km). Table 5 shows the remaining transport methods and distances. In this scenario, the manufacturing location of the underfloor heating system and heat collectors has not been changed.
- 50% Hybrid scenario assumes half of the energy required for heating has been produced by ASHP or GSHP and the remaining comes from NGB.
- 75% Hybrid scenario assumes 75% of heating energy has been provided by heat pumps and the remaining 25% produced by the gas boiler.

Table 5. Transport assumptions of manufacturing in Europe and Asia [36].

	Transport of;		Rail	Truck (Large)	Truck (Small)	Sea Freight
Transport (Europe)	raw materials to manufacturer	km	200	100		
	products from manufacturer to distributor	km	500	200		
	products for installation	km			200	
Transport (Asia)	raw materials to manufacturer	km	200	100		
	products from manufacturer to distributor	km		200		22,965
	products for installation	km			200	

The changes for these three scenarios are applied to both the baseline and the CE model to compare the impacts of these scenarios in the current year and an alternative of the year 2050.

3. Results

3.1. Baseline Results

The simulation results have been illustrated per functional unit, and the lifetime results are divided into the amount of total space heating demand for both heat pumps and gas boilers as the functional unit is decided as 'generating 1 kWh of thermal energy for domestic heating'. However, lifetime environmental impacts are also provided in the graphs to show the impact of technologies during their lifetime.

Environmental impacts of the baseline scenario for air source heat pump (ASHP), ground source heat pump (GSHP) and natural gas boiler (NGB) have been illustrated in Figure 2. ASHP has the highest impacts on average, and GSHP and NGB have 17% and 51% lower results than ASHP on average, respectively. When individual impact categories were investigated, the results illustrated that NGB has the lowest impact in all categories except Climate Change (CC)—(CC (Climate Change), OD (Ozone Depletion), TA (Terrestrial Acidification), FEU (Freshwater Eutrophication), MEU (Marine Eutrophication), HT (Human Toxicity), POF (Photochemical Oxidant Formation), PMF (Particulate Matter Formation), TE (Terrestrial Ecotoxicity), FE (Freshwater Ecotoxicity), ME (Marine Ecotoxicity), IR (Ionising Radiation), ALO (Agricultural Land Occupation), ULO (Urban Land Occupation), NLT (National Land Transformation), WD (Water Depletion), MD (Metal Depletion), FD (Fossil Depletion))—and Fossil Depletion (FD) categories.

Figure 2. Lifecycle environmental impacts of heat pumps and gas boilers for baseline scenario (HP: Heat pump, NGB: Gas boiler, UHS: Underfloor heating system, HC: Heat collector).

This study illustrates that emissions for ASHP and GSHP are reduced to 0.111 kg CO_2e/kWh and 0.097 kg CO_2e/kWh (Figure 2), respectively, compared with the literature [35], where there was a reduction from 0.276 kg CO_2e/kWh and 0.189 kg CO_2e/kWh. This is mainly because of the decarbonisation of the electricity mix through the high deployment of wind energy to replace coal and some part of natural gas during the last decade. The carbon intensity of the gas boiler is more than double both heat pumps with 0.241 kg CO_2e/kWh. NGB has 96.2 t CO_2e lifetime emissions, much higher than ASHP (42.3 t CO_2e) and GSHP (38.8 t CO_2e).

The two highest contributor phases of life cycle analysis are the 'use' and 'manufacturing' phases, which are responsible for 74% and 14% of all environmental impacts on average. The manufacturing of heating technologies, underfloor heating systems and heat collector phases accounts for 17%, 20% and 12% for ASHP, GSHP and NGB, respectively. It is important to keep in mind that the manufacturing of heat pumps occurs outside of the UK, which does not have an impact on the UK's territorial emissions; however, it will have an impact on consumption-based emissions of the UK or global emissions. The disposal phase accounts for 6%, 7% and 3% of total impacts for ASHP, GSHP and NGB, respectively; however, these impacts are negative due to contributions from the reuse of refrigerants and recycling of materials at the end of their life cycle. The refrigerant and maintenance phases account for only 3% for both heat pumps and no impact for the gas boiler as there is no refrigerant use in boilers. The transport phase, on the other hand, is only responsible for 1% of total environmental impacts.

When heat pumps are compared, GSHP has 17% lower results than ASHP as it requires less electricity because of its higher efficiency. The impact of heat collectors is relatively low in most of the categories, except the Terrestrial Acidification (TA), Photochemical Oxidant Formation (POF) and Particulate Matter Formation (PMF) categories. The reduction in the use phase in these categories is higher than the impact of manufacturing the heat collectors, so overall the environmental impact of GSHP remains lower than ASHP. The POF category is the only category in which GSHP has 3% higher results than ASHP because the impact of the manufacturing of heat collectors is greater than the reduction in the use phase. The highest difference between heat pumps occurs in the Ozone Depletion (OD) category with 36% because of lower refrigerant requirements. Metal Depletion (MD), Freshwater Eutrophication (FEU), Marine Eutrophication (MEU), Human Toxicity (HT) and Freshwater Ecotoxicity (FE) are the remaining categories that have more than 20% difference. Even though the disposal phase does not have a significant impact overall, there are several categories in which the disposal phase has higher impacts for heat pumps, such as TA, POF, PMF and ULO categories, accounting for 29%, 18%, 35%, and 22%, respectively.

3.2. Results from Future Scenarios

Scenario analysis aims to investigate the impact of changes planned in line with the government's targets and national policies. The Circular Economy (CE) scenario results expect the highest reductions for all heating technologies, and the Limited Growth (LG) scenario expects the lowest. Overall reductions in CE, RE and LG scenarios are 44%, 42% and 31% for heat pumps and 27%, 18% and 12% for the gas boiler (Figure 3).

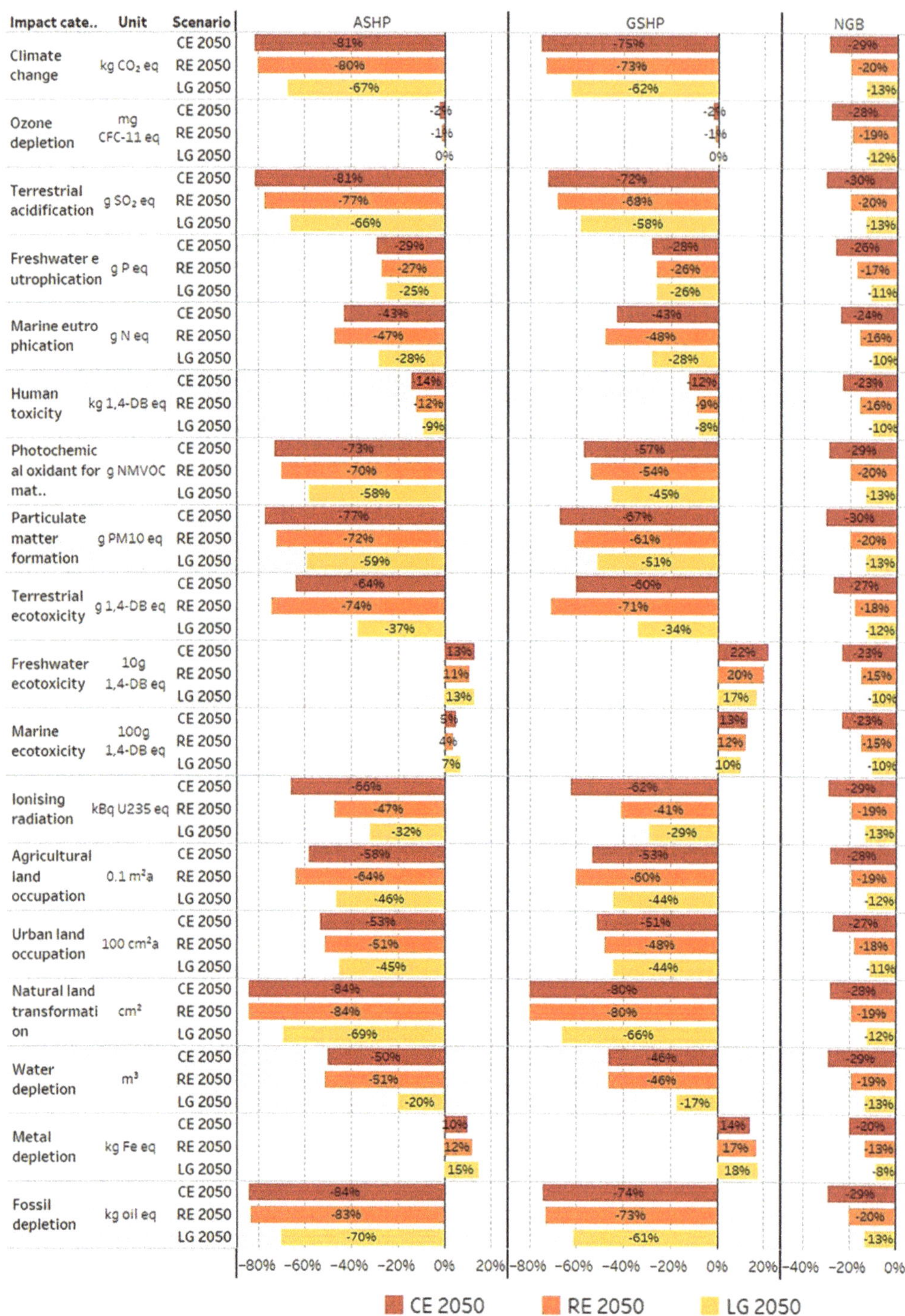

Figure 3. Lifecycle environmental impact change of future scenarios according to the baseline scenario.

In the CE scenario, the highest changes are in CC, TA, POF, PMF, NLT and FD categories with an average of 75% reduction in heat pumps. The lowest change occurs in the OD category with a 2% reduction only as the amount of refrigerant is the same in future scenarios. Even though the RE and LG scenario have lower results than the CE scenario, trends are the same. However, several categories expect an increase for all scenarios such as FE, ME, and MD. The main source of this impact is the heavy metals utilised in the high deployment of wind energy that will be provided by offshore wind farms; therefore, emissions to the water will be expected. Another toxicity category, human toxicity, also expects a lower reduction for all scenarios from 8% to 14% for heat pumps. Additionally, the major source of metal depletion comes from the life cycle of electricity because the high deployment of renewables requires more metal resources. On the other hand, there are several categories in which the RE scenario performs better than the CE scenario, such as MEU, TE, FE, ME and ALO. The main reason for this impact is that the CE scenario has the highest renewable share in the electricity mix and this has higher toxicity and land occupation results; however, the RE scenario has a lower renewable share and higher nuclear energy in the electricity mix. Therefore, negative impacts created by renewable energy are greater in the CE scenario. The LG scenario still has natural gas in the mix; therefore, the LG scenario still performs worse than both scenarios.

The reductions in NGB are very limited when compared with heat pumps. This is due to limited efficiency in the gas boiler. The reductions come from efficiency improvements in houses which will require less heat demand; therefore, the gas boiler expects similar reductions in all phases.

When the contributors to changes in future scenarios are investigated, only the use and disposal phases have an impact on categories. Figure 4 shows their weighted results, illustrating the importance of the use phase. Even though some categories are expecting significant increases in the disposal phase (ranging from 535–1286%), their weighted results are less than 1% when they are compared with the use phase. The highest disposal phase impact occurs in the OD category, with an increase of 20% and 9% for CE and RE scenarios and a decrease of 3% for the LG scenario.

3.3. *Transport and Hybrid Scenarios*

3.3.1. Results from Baseline Model

The Transport (SK) scenario results illustrate that changing the manufacturing location does not have a significant impact on most categories according to the baseline scenario. ASHP results show that even though the average change is less than 1%, there are some categories that have higher results (Figure 5). The highest impact category is MEU with a 30% decrease from the baseline scenario. TA, HT and PMF categories are other high impact categories with 13%, 11% and 8%, whereas with an increase, unlike the MEU category.

During life cycle phase analysis, only changes in the manufacturing of heat pumps, refrigerant and transport phases were considered. The manufacturing phase increases with an average of 27% in all categories, and the highest change occurs in the TE category with a 358% increase for ASHP (Figure 6). TA and PMF categories show increases of 226% and 58%, respectively. There are also several categories with negative impacts such as MEU and OD categories with a 92% and 19% decrease, respectively. The transport phase, on the other hand, increases 17% on average in all categories and the highest contribution comes from TA, PMF, MEU and POF categories with 77%, 49%, 40% and 39%, respectively. The refrigerant phase, however, has a negative impact, and results decrease only 2% on average and the highest change occurs in IR, NLT and TE categories with a decrease of 18%, 8% and 7%, respectively. PMF and TA categories have also seen a 4% increase.

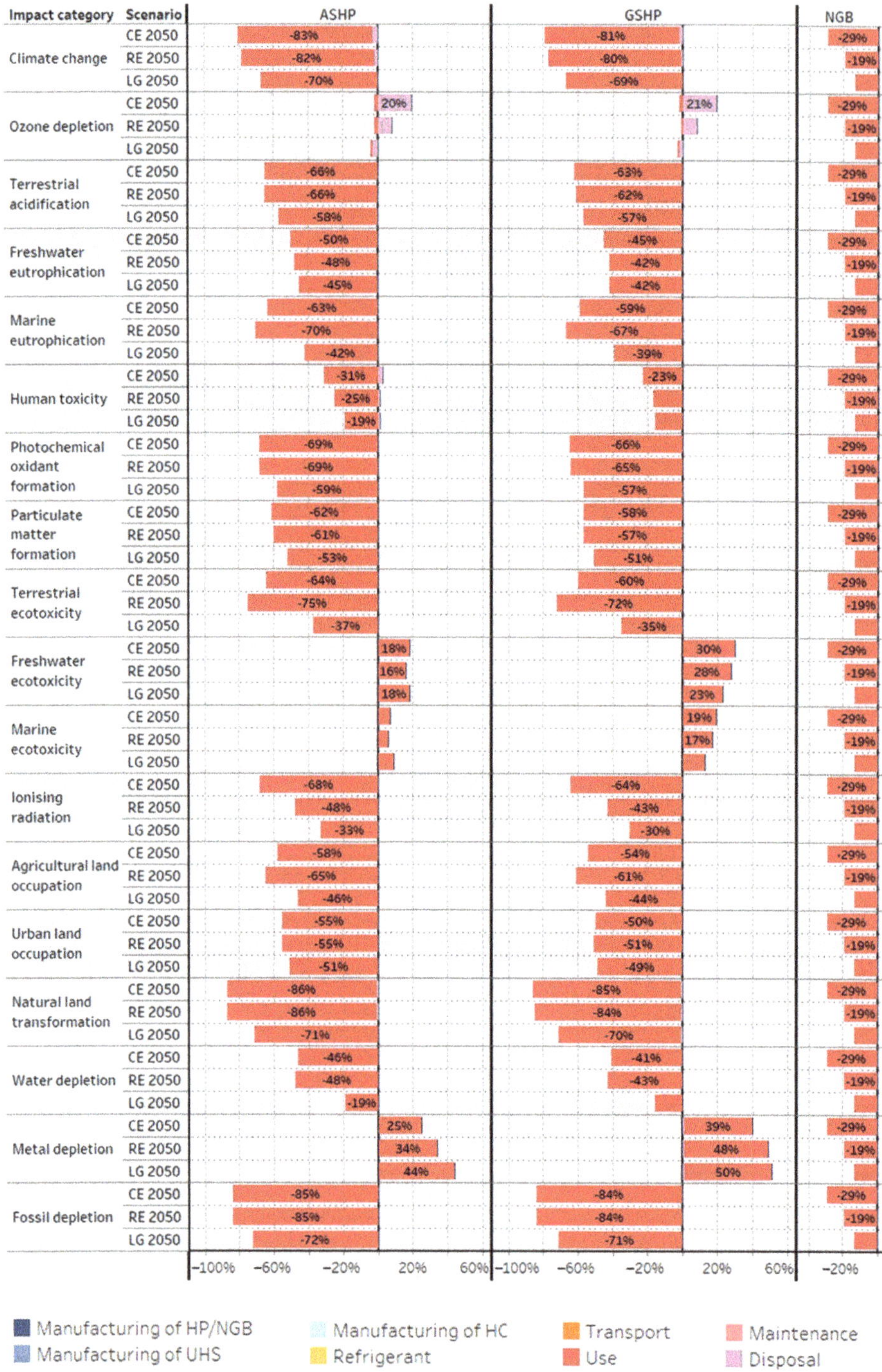

Figure 4. Lifecycle environmental impact change of phases in future scenarios.

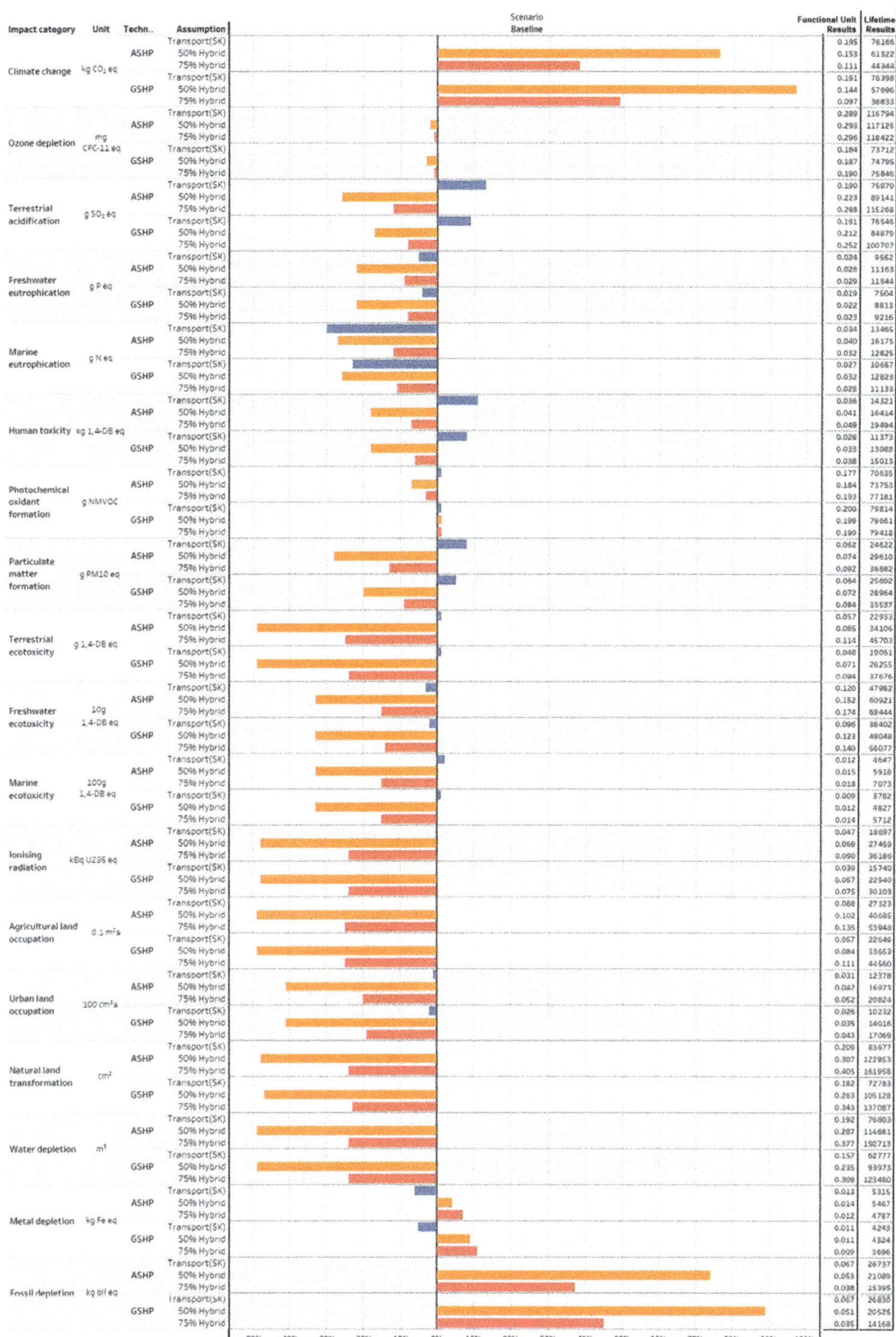

Figure 5. Lifecycle environmental impact change of Transport (SK) and Hybrid scenarios according to baseline scenario.

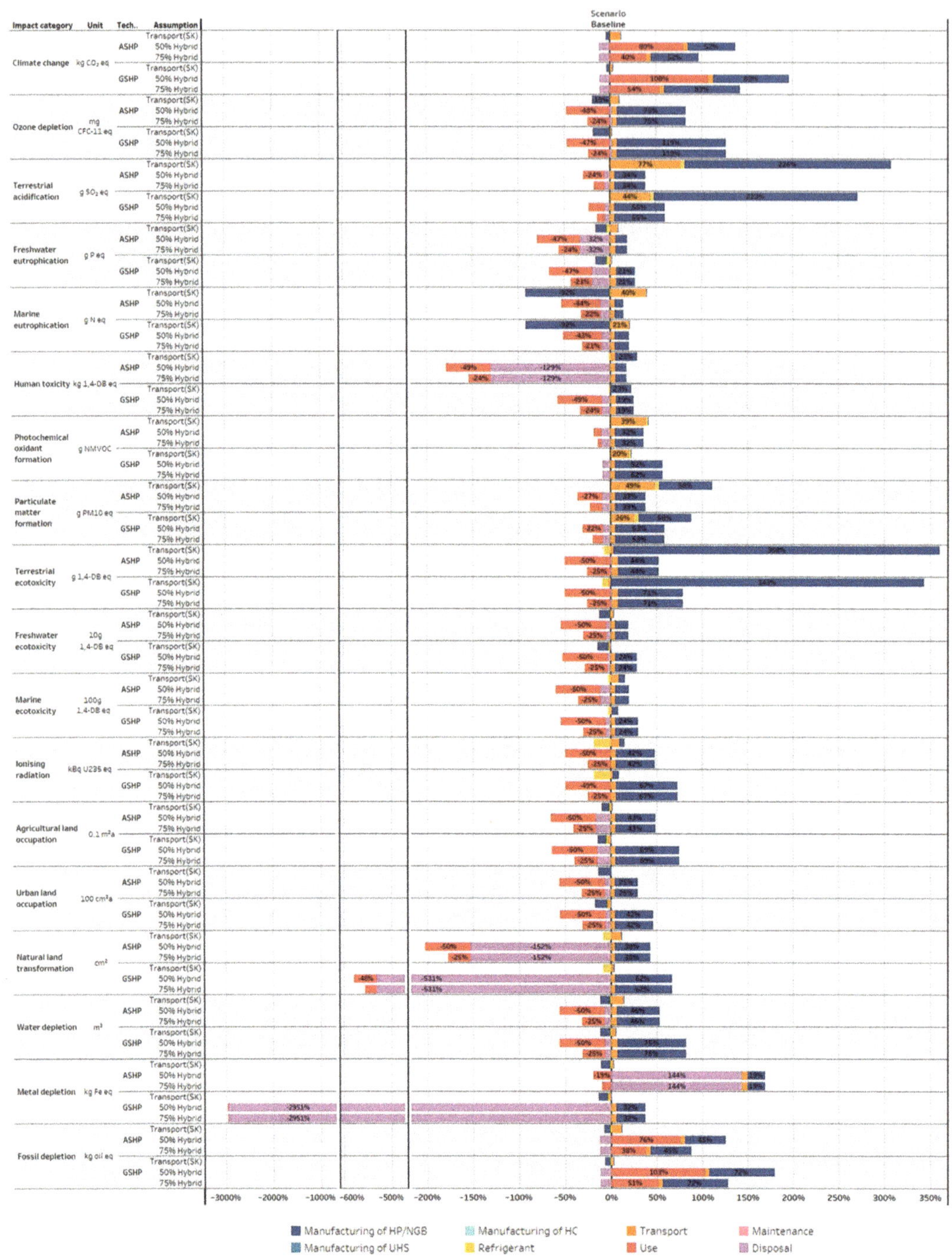

Figure 6. Lifecycle environmental impact change of phases for Transport (SK) and Hybrid scenarios according to baseline.

The results of GSHP show similarities with ASHP, with a decrease of 1% on average (Figure 5). The highest impact category is MEU with a 23% decrease, followed by a 9% and 8% increase in TA and HT categories, respectively. The changes in GSHP are relatively lower than ASHP as heat collectors in GSHP will still be manufactured in Europe in this scenario; therefore, the weight of the change becomes smaller in this technology.

The results of phases are also similar in manufacturing and refrigerants with a 26% increase and 2% decrease on average for GSHP (Figure 6). The highest impact categories are TE, TA, PMF and MEU categories in the manufacturing phase, and IR, TE, NLT, PMF and ULO categories in the refrigerant phase, like the ASHP results. The main difference between ASHP and GSHP occurs in the transport phase and the average change is 7%. Even though the highest categories are the same, the changes are less than ASHP.

The 50% Hybrid scenario results expect an increase of 32% and 20% on average in ASHP and GSHP, respectively (Figure 5). GSHP offers a lower increase or less reduction in all categories, resulting in fewer advantages than ASHP. The highest change occurs in CC and FD categories with a 76% and 79% increase for ASHP, and 97% and 89% increase for GSHP, respectively. The MD category also expects an increase of 3% and 7% for heat pumps. The remaining categories result in a decrease, and the highest decrease occurs in TE, IR, ALO, NLT, and WD categories, varying between 49% and 38% for both heat pumps. Some categories have a less than 5% impact change, such as OD, FEU and HT categories.

In the 50% Hybrid scenario, the highest changes occur in the disposal phase with an average of 15% and 200% decrease for ASHP and GSHP (Figure 6). Even though the overall change is greater in GSHP, most of the contribution comes from MD and NLT categories with a decrease of 2951% and 531%. The reason for this reduction is that the amount of metals required for ASHP is greater than GSHP; therefore, this value is a positive value for ASHP. Thus, negative metal depletion values coming from NGB reduce the impact of ASHP. When other phases are analysed, the use phase expects a decrease of 33% and 29% for ASHP and GSHP, and the manufacturing phase expects an increase of 33% and 53%, respectively. The transport phase has an average change of 5% increase for both heat pumps.

Even though the use phase offers a reduction in all categories, the CC and FD categories expect an increase in all phases except the disposal phase. As gas boilers perform worse than heat pumps only in this category in the baseline scenario (Figure 2), the hybrid scenario offers the worst results in these categories. Moreover, the MD category also expects an increase even though it is less than 10%. However, in other categories, the use phase eliminates the increases created by manufacturing and transport phases as the weight of the use phase is very large and creates negative results overall in all categories.

The 75% Hybrid scenario results offer less reduction than the half-hybrid scenario with an 11% and 9% decrease in ASHP and GSHP (Figure 5). Similarly, GSHP performs worse than ASHP in this scenario with an increase in CC and FD categories and a decrease in other categories; however, this scenario offers less decrease overall as the contribution of the gas boiler is less than the 50% Hybrid scenario. The highest changes occur in CC and FD categories with a 38% and 37% increase for ASHP, and 49% and 45% increase for GSHP, respectively. The highest decreases occur in TE, IR, ALO, and NLT categories, varying between 24% and 19% for both heat pumps.

3.3.2. Results from CE 2050 Model

The Transport (SK) scenario results show that changing the manufacturing location could increase the environmental impacts on average 3% and 1% for ASHP and GSHP, respectively, according to the CE 2050 model (Figure 7). The highest changes for ASHP occur in TA and PMF with a 68% and 34% increase. Additionally, results suggest a decrease in several categories with less than 3% except the MEU category, which has a 53% reduction in the CE 2050 model. GSHP results show lower values than ASHP in all categories, but the highest contributors are the same impact categories.

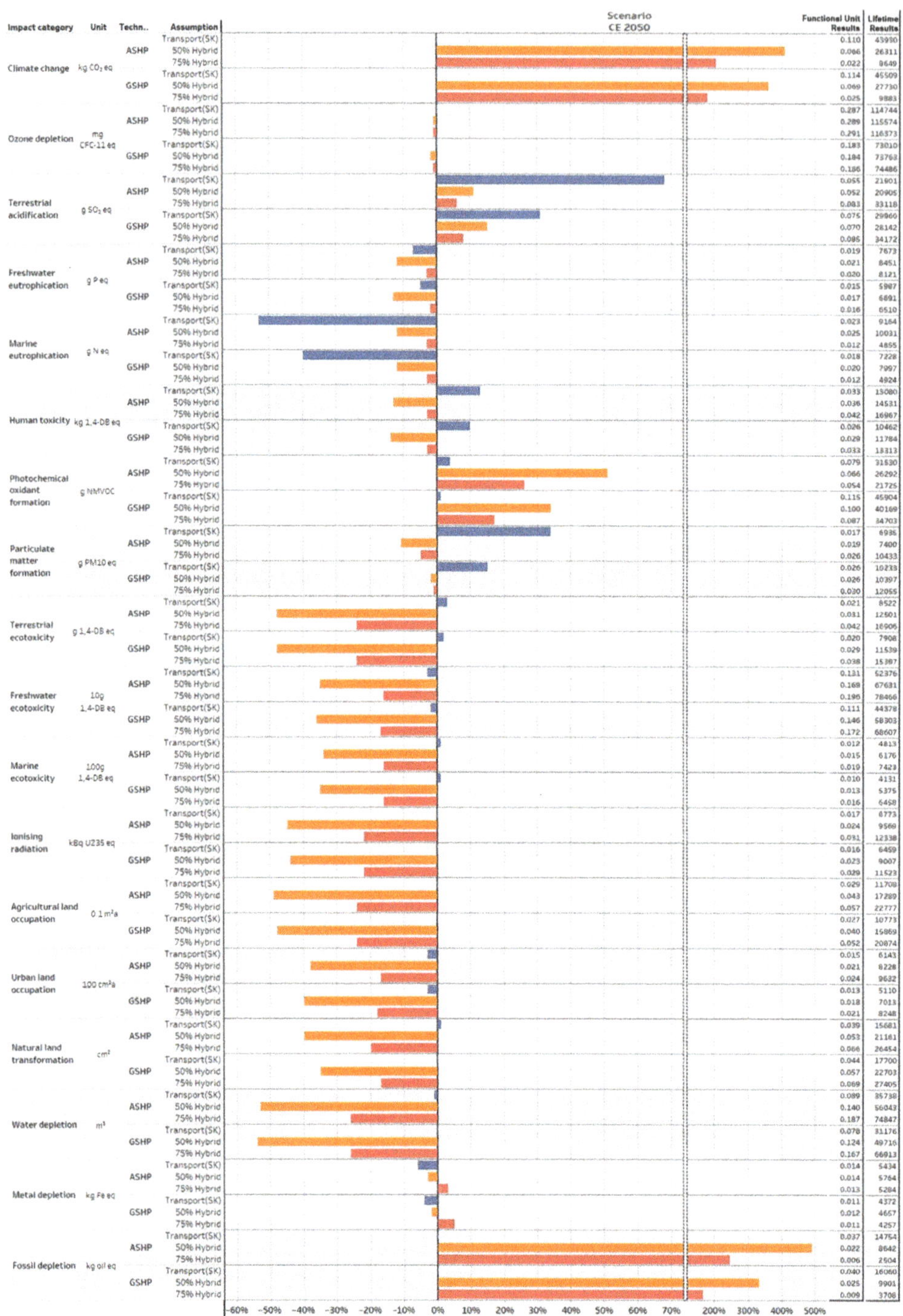

Figure 7. Lifecycle environmental impact change of Transport (SK) and Hybrid scenarios according to CE scenario.

The life cycle phase results illustrate that the highest contributor phases to the changes from the CE 2050 model are the manufacturing of heat pumps, refrigerant, and transport phases, similar to the baseline model (Figure 8). The results of changes in these phases are the same with the baseline model; therefore, the changes in these phases have the same impacts in both the baseline and CE 2050 model.

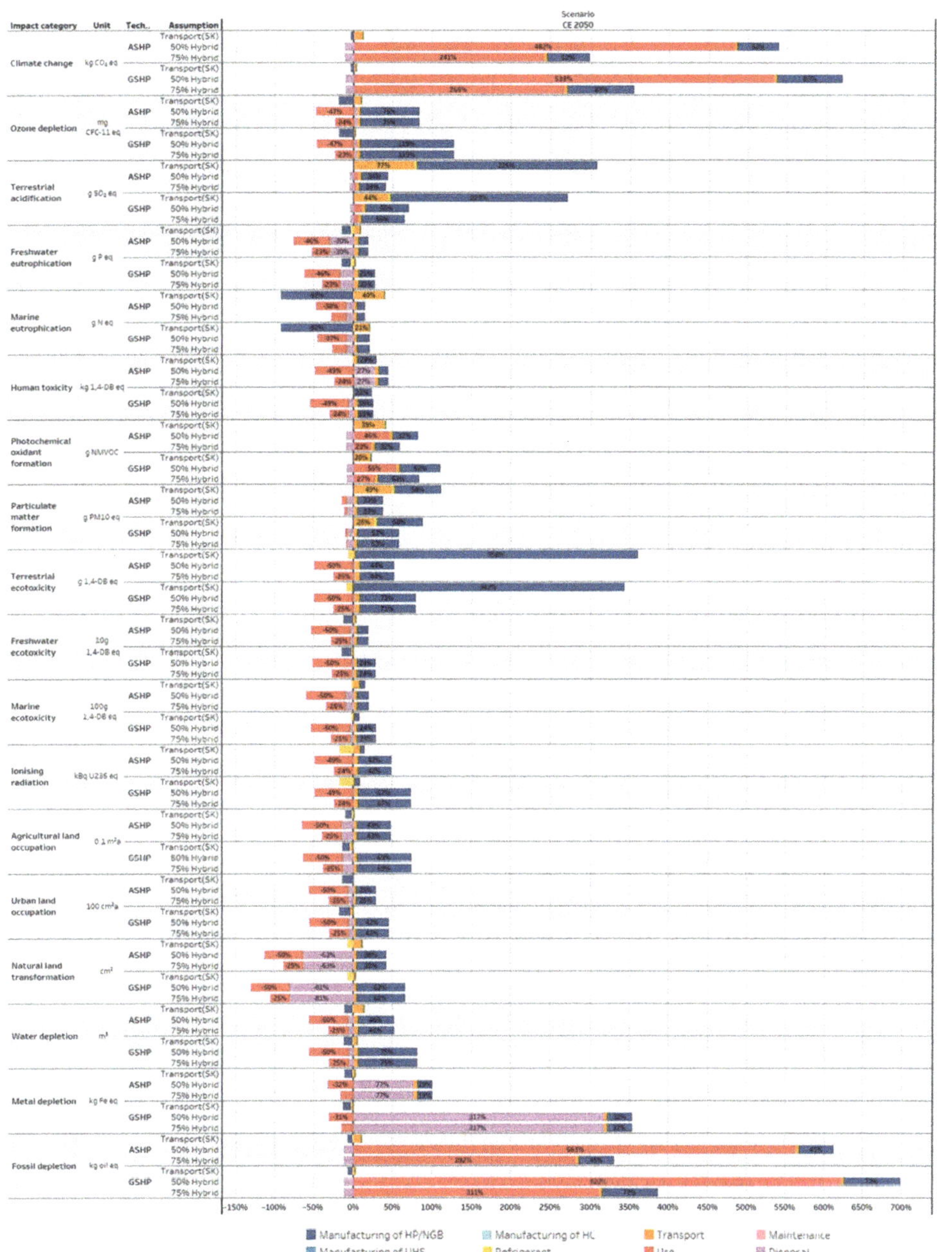

Figure 8. Lifecycle environmental impact change of phases for Transport (SK) and Hybrid scenarios according to CE scenario.

Even though hybrid scenarios in the CE 2050 model have similar results as the baseline model in most of the categories, there is a significant difference in CC and FD categories as they are very sensitive to the use phase results. In the 50% Hybrid scenario, the highest changes occur in the FD category with a 490% and 333% increase for ASHP and GSHP, similar to the baseline model (Figure 7). The other category suggesting an increase is CC with 409% and 360% for both heat pumps. The impact of the MD category is lower than the baseline model in the CE model. Most of the remaining categories have a reduction of around 16–47%.

The results of phases in the 50% Hybrid scenario illustrate that the highest changes occur in the manufacturing phase, with a 33% and 53% increase on average for both heat pumps (Figure 8). The transport phase creates an increase of 5% and 6% for ASHP and GSHP, respectively. The disposal phase, on the other hand, expects a decrease of 4% for ASHP and an increase of 8% for GSHP. However, the use phase suggests a decrease of around 5% on average for both heat pumps. Similar to the baseline model, the use phase offers a reduction in all categories and an increase for CC and FD categories with 482% and 563% for ASHP, and 533% and 622% for GSHP, respectively. The only exception is for the POF category, which was expecting a reduction in the baseline model but expecting an increase in the CE 2050 model. The main reason for this is that the result of NGB for this category is lower than heat pumps in the baseline model; however, in the CE 2050 model, NGB has a higher value and increases the average of hybrid results.

TA, FEU, and PMF categories have a reduction varying between 9% and 18%, whereas the remaining categories expect higher reductions varying between 30% and 48%. In the CE 2050 model, hybrid scenarios offer an overall increase in contrast to the baseline model mainly because the change in the CC category is greater than the baseline model and the weight of the use phase is lower in the CE 2050 model.

Similar to the baseline model, the 75% Hybrid scenario results offer less increase than the half-hybrid scenario with a 4% and 10% increase overall in ASHP and GSHP, respectively. The highest change occurs in CC and FD categories with a 205% and 246% increase for ASHP, and 181% and 167% increase for GSHP, respectively. The highest decreases occur in TE, IR, ALO and WD categories, varying between 19% and 22% for both heat pumps.

The changes in manufacturing, transport and disposal phases are similar to the baseline model in both hybrid scenarios, so there is no difference between the baseline and CE model and 50% and 75% Hybrid scenarios in these phases, except the use phase.

The results of hybrid scenarios offer a benefit to reduce the negative impacts caused by heat pumps in most of the categories. Even though this creates an increase in CC and FD categories and GHG emissions, negative consequences could be prevented. Moreover, replacing gas boilers with heat pumps requires a transition period, and hybrid applications could help to create a smoother transition.

3.4. Data Quality and Limitations

In order to validate the study, results are compared with the adopted study [35]. Impact categories vary between different calculation methods, but several impact categories are common in most of the studies so only these categories are compared. The CC impact result of ASHP is 0.225 kg CO_2e/kWh in the baseline model, which is 18% lower than the adopted study (0.276 kg CO_2e/kWh). The GSHP result is 0.168 kg CO_2e/kWh for the baseline model and the result from the adopted study is 0.189 kg CO_2e/kWh, which is lower around 11%. The OD category of the adopted study was 0.3 mg R11eq, which is 2% higher than this study (0.294 mg CFC-11eq). Additionally, TA category results for ASHP and GSHP were 0.86 and 0.59 g SO2eq, which is 2% and 8% lower than this study's results, respectively (0.842 and 0.638 g SO2eq). FEU and HT categories have higher differences that vary between 20% and 47%. The major reason causing these differences is the different methodology used for the models. This study used ReCiPe Midpoint (H) methodology;

however, the adopted study used CML 2 Baseline 2001 methodology. Moreover, the adopted study used GaBi software, and this study used SimaPro software.

The limitation of the Transport (SK) scenario is that even though South Korea is used as a manufacturing location, rest-of-the-world (RoW) data for production assumptions and input data have been used in SimaPro due to the lack of data availability. Transport simulations are specific to South Korea; however, manufacturing data are not specific.

The impacts of the electricity mix, heat demand, efficiencies of technologies, lifetime of the products and disposal phase have been assessed for a sensitivity analysis. The parameters have been decided as:

- Doubling renewable share in the electricity mix;
- 50% increase in SPF (in this analysis, the efficiency of the gas boiler has been increased from 90% to 95%);
- 25% reduction in heat demand;
- 25% increase in product lifetimes;
- 25% increase in recycling rates of materials.

The results of sensitivity analysis indicate that electricity use has a significant impact on heat pump results. Doubling the renewable share in the electricity mix creates positive and negative impacts in several categories for ASHP (Figure 9). The highest influences occur on IR, NLT, FD, and CC categories with a decrease of around 41%, 41%, 40% and 34%, respectively. However, it could increase the results of TE, ALO, WD, FE, MEU and ME categories with 97%, 95%, 76%, 52%, 42% and 42%. The renewable share has no impact on NGB as it uses natural gas only.

A 50% increase in SPF creates an average of 29% reduction overall, and the highest changes occur in TE and ALO categories, accounting for 70% and 50%, respectively. The remaining categories expect a reduction range from 8% to 39%. Increasing boiler efficiency from 90% to 95% reduces all impact categories by an average of 4%.

A 25% reduction in demand has both negative and positive impacts on categories. Even though the lifetime results expect a reduction in this analysis, functional unit results fluctuate as the lifetime results are divided into heat demand, which is 25% reduced. Therefore, some categories react differently in lifetime and functional unit results. The highest changes occur in TE and ALO categories, similar to SPF improvements for heat pumps. A similar issue occurs for the gas boiler and creates an increase of 4% overall, even though lifetime results are reduced.

Increasing the lifetime of products to 25 years increases the lifetime impact results as expected, with an increase of an average 16%, 15% and 22% for ASHP, GSHP and NGB. However, functional unit results expect a decrease of 7%, 8% and 2% for the technologies, respectively.

A 25% increase in the recycling rates of materials also has a significant impact in several categories for heat pumps such as TE, MEU and ALO categories, with a reduction of 56%, 26% and 25%, and the WD category with an increase of 36% for heat pumps. However, its impact is relatively low for gas boilers.

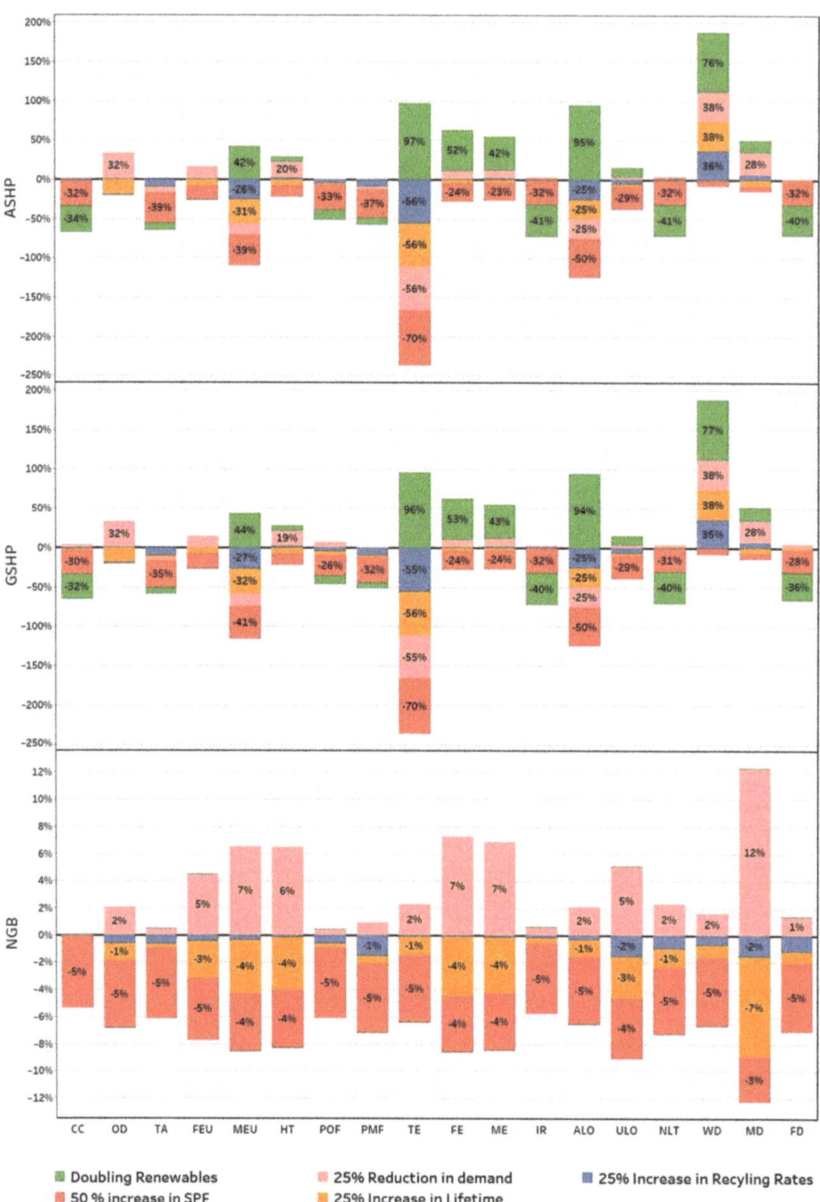

Figure 9. Impacts of different parameters on environmental results for heating technologies in sensitivity analysis.

4. Conclusions

This study assesses the environmental impacts of heat pumps vs. gas boilers through three main scenarios: Circular Economy (CE), Resource Efficiency (RE) and Limited Growth (LG), and three alternative scenarios: Transport (SK), 50% Hybrid and 75% Hybrid. The findings illustrate that replacing gas boilers with heat pumps could help to reduce lifetime GHG emissions by 78% (CE scenario), 77% (RE scenario) and 65% (LG scenario). The overall average impact is expected to be lower around 43% (CE scenario), 42% (RE scenario) and 31% (LG scenario). However, the following categories MEU, TE, FE, ME, ALO and WD perform 5% lower in the CE than in the RE scenario.

Heat pumps provide significant reductions in GHG emissions and the fossil depletion category; however, they do not provide sustainable solutions in other impact categories. Moreover, future scenarios expect reductions in most of the categories; however, several categories expect an increase in contrast to remaining impact categories in all scenarios, such as freshwater ecotoxicity, marine ecotoxicity and metal depletion categories. It is important to point out that the high deployment of renewables, especially offshore wind farms, will have a positive impact in most of the categories, but also create toxicity problems and material scarcities.

Hybrid scenario results (50% Hybrid and 75% Hybrid) expect an increase in GHG emissions as boilers use fossil fuel, whereas the negative impacts coming from the remaining categories decrease. Therefore, a transition period that includes hybrid applications rather than replacing gas boilers individually should be provided in order to reduce negative impacts. In both hybrid scenarios, the overall results suggest a reduction in the baseline model (22% for 50% Hybrid scenario and 10% for 75% Hybrid scenario); however, the changes are 15% lower in the CE scenario. In the CC category, the changes are greater in the CE model as heat demand in the future will be relatively small; therefore, the importance of each phase will be higher to reduce the negative impacts. As the UK increases its ambitions to reach the 'Net-Zero' target, actions for each phase should be considered thoroughly.

In the Transport (SK) scenario, changing the manufacturing location from Europe to Asia creates a 1% reduction in the baseline model and a 2% increase in the CE model. The reason for this slight increase is that the weight of the use phase is lower in the CE scenario due to efficiency improvements in houses and low-carbon technologies, so the remaining phases comprise higher shares. As the main contributor to these changes is the manufacturing phase, better production lines through adapting CE principles could help to reduce the impact of the manufacturing phase. It is also important to reiterate that, even though the impact of the manufacturing phase is relatively smaller than the remaining phases (14% of the overall impact), the manufacturing of heat pumps has an impact in those locations where manufacturing takes place; therefore, this does not count in territorial emissions.

Future scenarios show how decision making could have a significant impact on environmental impacts. The CE scenario provides the best outcome among all scenarios without affecting economic growth. Reducing GHG emissions and preventing negative consequences are highlighted in the CE scenario. Achieving the Net-Zero target requires strong commitments, and the results of future scenarios emphasise that the importance of impacts proposed by changes will reduce in time. Therefore, quick implementation of changes and stronger commitments are required in other areas as well, mainly energy efficiency improvement in houses (insulation, etc.), better-installed heat pumps with higher efficiencies and greener manufacturing solutions.

High demand for specific materials could enhance scarcities and environmental degradation related to resource extraction and processing. Circular economy principles through reuse and recycle options become more important in these situations. However, new strategies are needed to reach the 'Net-Zero' target as it requires stronger commitments and more rapid market dissemination. Therefore, a comprehensive approach through a market introduction programme should be provided at the beginning before shifting from one technology to another. It is important to stress that different heating technologies require different material demands and waste streams. High deployment of heat pumps in the CE scenario (17.7 million) will require high demand for metals and minerals, even though they do not have significant impacts on GHG emissions in the manufacturing phase. It would be of utmost importance to develop CE standards for the production of heat pumps, e.g., through procurement or eco-design, and include the use of secondary materials and the re-usability of all components. Thus, a more comprehensive circular framework for decision-making tools could be created for sustainable design practice. A holistic approach should be considered where both territorial and consumption-based emissions are considered together for policies and future planning.

Author Contributions: Conceptualization and methodology, S.S., C.S. and T.D.A.; software, S.S., T.D.A.; data collection, analysis and validation, S.S.; visualization, S.S.; writing—original draft preparation, S.S., C.S.; writing—review and editing, S.S., C.S., T.D.A., R.B.; supervision, C.S. All authors have read and agreed to the published version of the manuscript.

Funding: This research received no external funding.

Institutional Review Board Statement: Not applicable.

Informed Consent Statement: Not applicable.

Data Availability Statement: Publicly available datasets were analyzed in this study.

Conflicts of Interest: The authors declare no conflict of interest.

Abbreviations

ALO	Agricultural land occupation	MEU	Marine eutrophication
ASHP	Air source heat pump	NGB	Natural gas boiler
CC	Climate change	NLT	Natural land transformation
CCC	Climate Change Committee	OD	Ozone depletion
CE	Circular Economy	PMF	Particulate matter formation
FD	Fossil depletion	POF	Photochemical oxidant formation
FE	Freshwater ecotoxicity	RE	Resource Efficiency
FEU	Freshwater eutrophication	RHI	Renewable Heat Incentive
GSHP	Ground Source Heat Pump	RIBA	Royal Institute of British Architects
HC	Heat Collector	SK	South Korea
HP	Heat Pump	SPF	Seasonal performance factor
HT	Human toxicity	TA	Terrestrial acidification
IR	Ionising radiation	TE	Terrestrial ecotoxicity
LCA	Life cycle assessment	UHS	Underfloor heating system
LG	Limited growth	UK	United Kingdom
MD	Metal depletion	ULO	Urban land occupation
ME	Marine ecotoxicity	WD	Water depletion

References

1. The World Bank Urban Development Overview. Available online: http://www.worldbank.org/en/topic/urbandevelopment/overview (accessed on 14 November 2020).
2. UN-Habitat Energy. Available online: https://unhabitat.org/urban-themes/energy/ (accessed on 14 November 2018).
3. UNEP The New Urban Agenda—Habitat III. Available online: http://habitat3.org/the-new-urban-agenda (accessed on 26 December 2018).
4. World Green Building Council Global Status Report 2017. Available online: https://www.worldgbc.org/news-media/global-status-report-2017 (accessed on 28 April 2020).
5. CCC. *Net Zero-The UK's Contribution to Stopping Global Warming*; The Climate Change Committee: London, UK, 2019; Volume 277.
6. HM Government. *The Clean Growth Strategy—Leading the Way to a Low Carbon Future*; Department for Business, Energy & Industrial Strategy: London, UK, 2017.
7. Gaujena, B.; Borodinecs, A.; Zemitis, J.; Prozuments, A. Influence of building envelope thermal mass on heating design temperature. *IOP Conf. Ser. Mater. Sci. Eng.* **2015**, *96*. [CrossRef]
8. Ovchinnikov, P.; Borodiņecs, A.; Millers, R. Utilization potential of low temperature hydronic space heating systems in Russia. *J. Build. Eng.* **2017**, *13*, 1–10. [CrossRef]
9. Valancius, R.; Singh, R.M.; Jurelionis, A.; Vaiciunas, J. A review of heat pump systems and applications in cold climates: Evidence from Lithuania. *Energies* **2019**, *12*, 4331. [CrossRef]
10. Yıldız, A.; Yıldırım, R. Investigation of using R134a, R1234yf and R513A as refrigerant in a heat pump. *Int. J. Environ. Sci. Technol.* **2020**. [CrossRef]
11. BEIS. *Heat Pump Manufacturing Supply Chain Research Project*; Department for Business, Energy & Industrial Strategy: London, UK, 2020.

12. Zajacs, A.; Lalovs, A.; Borodinecs, A.; Bogdanovics, R. Integration of Renewable Energy in the Built Environment (Electricity, Heating and Cooling)—Small ammonia heat pumps for space and hot tap water heating. *Energy Procedia* **2017**, *122*, 74–79. [CrossRef]
13. BEIS. *Future Support for Low Carbon Heat: Consultation*; Department for Business, Energy & Industrial Strategy: London, UK, 2020.
14. EUROSTAT. Energy for Heating/Cooling from Renewable-2017. Available online: https://ec.europa.eu/eurostat/web/products-eurostat-news/-/DDN-20190304-1?inheritRedirect=true&redirect=%2Feurostat%2F (accessed on 28 January 2020).
15. Etude. *Low Carbon Heat: Heat Pumps in London*; Greater London Authority: London, UK, 2018.
16. CCC. *Reducing UK emissions—2019 Progress Report to Parliament*; The Climate Change Committee: London, UK, 2019; Volume 93.
17. National Grid. *Future Energy Scenarios 2019*; National Grid: London, UK, 2019.
18. CCC. *UK Housing: Fit for the Future?* The Climate Change Committee: London, UK, 2019.
19. Ofgem. Domestic RHI Quarterly Report Issue 23. Available online: https://www.ofgem.gov.uk/publications-and-updates/domestic-renewable-heat-incentive-quarterly-report-issue-23 (accessed on 28 April 2020).
20. CCC. *An Independent Assessment of the UK's Clean Growth Strategy from Ambition to Action*; Department for Business, Energy & Industrial Strategy: London, UK, 2018.
21. HM Government. *The Ten Point Plan for a Green Industrial Revolution*; Department for Business, Energy & Industrial Strategy: London, UK, 2020.
22. MHCLG. *The Future Homes Standard—2019 Consultation on Changes to Part L (Conservation of Fuel and Power) and Part F (Ventilation) of the Building Regulations for New Dwellings*; Ministry of Housing, Communities and Local Government: London, UK, 2019. Available online: https://assets.publishing.service.gov.uk/government/uploads/system/uploads/attachment_data/file/852605/Future_Homes_Standard_2019_Consultation.pdf (accessed on 20 April 2020).
23. MHCLG. *The Future Home Standard: 2019—Consultation on changes to Part L (Conservation of Fuel and Power) and Part F (Ventilation) of the Building Regulations for New Dwellings—Summary of Responses Received and Government Response*; Ministry of Housing, Communities and Local Government: London, UK, 2021. Available online: https://assets.publishing.service.gov.uk/government/uploads/system/uploads/attachment_data/file/956094/Government_response_to_Future_Homes_Standard_consultation.pdf (accessed on 20 February 2021).
24. BEIS Heat in Buildings. Available online: https://www.gov.uk/government/groups/heat-in-buildings (accessed on 19 October 2020).
25. RIBA. *RIBA 2030 Climate Challenge*; Royal Institute of British Architects: London, UK, 2019.
26. DECC. *Potential Cost Reductions for Ground Source Heat Pumps: The Scope for a Mass Market*; Department of Energy & Climate Change: London, UK, 2016.
27. DECC. *Potential Cost Reductions for Air Source Heat Pumps: The Scope for a Mass Market*; Department of Energy & Climate Change: London, UK, 2016.
28. Sevindik, S.; Spataru, C.; Domenech Aparisi, T.; Bleischwitz, R. Comparative Environmental Impact Assessment of Heat Pumps with Gas Boilers and Scenario Analysis Towards a Circular Economy in the UK. In Proceedings of the 15th Conference on Sustainable Development of Energy, Water and Environment Systems (SDEWES), Cologne, Germany, 1–5 September 2020; Ban, M., Duić, N., Schneider, D.R., Buonomano, A., Eds.; Faculty of Mechanical Engineering and Naval Architecture: Zagreb, Croatia, 2020.
29. Sevindik, S.; Spataru, C. Life Cycle Assessment of Domestic Heat Pumps with Gas Boiler and Hybrid Scenario Analysis in the UK. In Proceedings of the International Conference on Energy, Environment and Storage of Energy (ICEESEN)), Conference Proceedings, Kayseri, Turkey, 19–21 November 2020; Akansu, S.O., Unalan, S., Eds.; 2020; ISBN 978-625-409-147-6.
30. Curran, M.A. Life Cycle Assessment: A review of the methodology and its application to sustainability. *Curr. Opin. Chem. Eng.* **2013**, *2*, 273–277. [CrossRef]
31. ISO BSI EN ISO 14040:2006 Environmental Management—Life Cycle Assessment—Principles and Framework. Available online: https://www.iso.org/obp/ui/#iso:std:iso:14040:ed-2:v1:en (accessed on 8 January 2019).
32. ISO. ISO 14044:2006: Environmental Management-Life Cycle Assessment-Requirements and Guidelines, International Organization for Standardization 2006. Available online: https://www.iso.org/obp/ui/#iso:std:iso:14044:ed-1:v1:en (accessed on 20 February 2020).
33. PRé Sustainability SimaPro 8.0.3 LCA Software 2014. Available online: https://pre-sustainability.com (accessed on 5 May 2019).
34. Huijbregts, M.A.J.; Steinmann, Z.J.N.; Elshout, P.M.F.; Stam, G.; Verones, F.; Vieira, M.; Zijp, M.; Hollander, A.; van Zelm, R. ReCiPe2016: A harmonised life cycle impact assessment method at midpoint and endpoint level. *Int. J. Life Cycle Assess.* **2017**, *22*, 138–147. [CrossRef]
35. Greening, B.; Azapagic, A. Domestic heat pumps: Life cycle environmental impacts and potential implications for the UK. *Energy* **2012**, *39*, 205–217. [CrossRef]
36. Wernet, G.; Bauer, C.; Steubing, B.; Reinhard, J.; Moreno-Ruiz, E.; Weidema, B. The Ecoinvent Database Version 3 (Part I): Overview and Methodology. Available online: https://www.ecoinvent.org/ (accessed on 27 April 2020).
37. Passarini, F.; Ciacci, L.; Nuss, P.; Manfredi, S. *Material Flow Analysis of Aluminium, Copper, and Iron in the EU-28*; Joint Research Centre: Ispra, Italy, 2018; ISBN 9789279857447.
38. Environment Agency Waste Electrical and Electronic Equipment (WEEE): Evidence and National Protocols Guidance. Available online: https://www.gov.uk/government/publications/weee-evidence-and-national-protocols-guidance/waste-electrical-and-electronic-equipment-weee-evidence-and-national-protocols-guidance (accessed on 10 December 2020).

39. BEIS RHI Monthly Deployment Data: October 2020. Available online: https://www.gov.uk/government/statistics/rhi-monthly-deployment-data-october-2020 (accessed on 8 December 2020).
40. CCC. Sectoral scenarios for the fifth carbon budget—Technical report. *Carbon Budg.* **2015**. [CrossRef]
41. DUKES Electricity Statistics—GOV.UK. Available online: https://www.gov.uk/government/collections/electricity-statistics (accessed on 28 April 2020).
42. BEIS New Plans to Make UK World Leader in Green Energy. Available online: https://www.gov.uk/government/news/new-plans-to-make-uk-world-leader-in-green-energy (accessed on 30 March 2021).
43. ETI. *Pathways to Low Carbon Heating: Dynamic Modelling of Five UK Home*; Energy Technologies Institute: Birmingham, UK, 2019; Volume 44.

Article

Towards a Digital Product Passport Fit for Contributing to a Circular Economy

Thomas Adisorn *, **Lena Tholen** and **Thomas Götz**

Energy, Transport and Climate Policy Division, Wuppertal Institute, 42103 Wuppertal, Germany; lena.tholen@wupperinst.org (L.T.); thomas.goetz@wupperinst.org (T.G.)
* Correspondence: thomas.adisorn@wupperinst.org; Tel.: +49-202-2492-246

Abstract: The Digital Product Passport (DPP) is a concept of a policy instrument particularly pushed by policy circles to contribute to a circular economy. The preliminary design of the DPP is supposed to have product-related information compiled mainly by manufactures and, thus, to provide the basis for more circular products. Given the lack of scientific debate on the DPP, this study seeks to work out design options of the DPP and how these options might benefit stakeholders in a product's value chain. In so doing, we introduce the concept of the DPP and, then, describe the existing regime of regulated and voluntary product information tools focusing on the role of stakeholders. These initial results are reflected in an actor-centered analysis on potential advantages gained through the DPP. Data is generated through desk research and a stakeholder workshop. In particular, by having explored the role the DPP for different actors, we find substantial demand for further research on a variety of issues, for instance, on how to reduce red tape and increase incentives for manufacturers to deliver certain information and on how or through what data collection tool (e.g., database) relevant data can be compiled and how such data is provided to which stakeholder group. We call upon other researchers to close the research gaps explored in this paper also to provide better policy direction on the DPP.

Keywords: resource efficiency; product policy; energy efficiency; digitalization; life cycle assessment; easy-to-repair design

Citation: Adisorn, T.; Tholen, L.; Götz, T. Towards a Digital Product Passport Fit for Contributing to a Circular Economy. *Energies* **2021**, *14*, 2289. https://doi.org/10.3390/en14082289

Academic Editor: Anna Mazzi

Received: 1 April 2021
Accepted: 15 April 2021
Published: 19 April 2021

Publisher's Note: MDPI stays neutral with regard to jurisdictional claims in published maps and institutional affiliations.

Copyright: © 2021 by the authors. Licensee MDPI, Basel, Switzerland. This article is an open access article distributed under the terms and conditions of the Creative Commons Attribution (CC BY) license (https://creativecommons.org/licenses/by/4.0/).

1. Introduction

At the international level, with the Agenda 2030 [1] the global community has defined 17 Sustainable Development Goals (SDGs) for socially, economically and ecologically sustainable development [2]. Sustainable development in general and the SDGs in particular require suitable indicators and corresponding data in order to initiate necessary policy action and to measure progress.

On the level of the European Union (EU) and with regard to product policy, the provision of data and the organization of a comprehensive information flow is promoted, among other things, by the "European Green Deal" [3] and the "Circular Economy Action Plan" [4] of the EU. Another impetus that makes the topic of product policy and data collection/provision even more relevant is the topic of digitalization, which has been heavily discussed for years (cf. [5]). In this context, a concept that is gaining attention in the political agenda is the development of a Digital Product Passport (DPP), which is not only topic in the two already mentioned EU strategies but also confirmed in the "Council conclusions on Making the Recovery Circular and Green" drafted under the German EU Council Presidency [6]. For providing input to the German Council Presidency of the second half of 2020, the authors of this article developed a scoping paper on the DPP, which this article is based on [7]. From the anchoring in high-level policy strategies, one can derive the high expectations on the DPP as an essential new tool for enabling a holistic and comprehensive recording of sustainability aspects in the future. Among other things, the DPP is intended to provide consistent "track and trace" information on the origin, composition, repair and dismantling options of a product, as well as on its handling at the end of its service life. The aim of the DPP is not only to promote a circular economy and thus support a low-carbon transition but also to overcome existing obstacles like the lack of information. The DPP has the potential to provide different actors (such as consumers and waste management companies) with relevant information on a product and thus force decisions towards sustainable development (for consumers during the purchase and use phase, for waste management companies during disassembling and recycling). For this undertaking, e.g., Gligoric et al. have been developing smart tags based on printed sensors to product or object identification on a per item-level [8], while Donetskaya and Gatchin in their conference paper come up with some requirements for the content of a DPP [9]. Depending on its exact design, it may help companies along the value chain to develop sustainable business models. For instance, Longo et al. argue to manufacture batteries and vehicles "with fewer, renewable, recyclable/recycled, and non-hazardous materials and characterized by lower energy and environmental impacts during their life cycle" [10] and Wielgosiński et al. call for a reduction of waste streams by having raw materials circulated in the domestic market [11]. To make businesses deliver to these objectives, the obligation to generate high quality product information can be a valuable contribution in a policy mix for an effective circular approach [12].

At the European level, the DPP is most prominently discussed in the context of the Sustainable Products Initiative (SPI) [13] in combination with the expansion of the EU Ecodesign Directive beyond energy-related products to include as wide a range of products as possible in order to define appropriate minimum sustainability and information requirements for specific product groups. Following this, DPP and SPI are also closely related to other recent EU initiatives such as in particular "Consumer policy-strengthening the role of consumers in the green transition" [14]. The central objective of the latter is to revise EU policy within the framework of the "European Consumer Agenda" [15], to enable consumers to play a more active role in the timely transition to a more sustainable economy ("green transition") by providing reliable and useful product information. Among other things, minimum requirements for sustainability logos and quality labels as well as reliable environmental information, e.g., on service life and repair options, are to prevent claims from being glossed over in the sense of "greenwashing" (i.e., giving a false impression of the actual environmental impact) or products being sold with a shortened service life. In addition, as part of the EU initiative "Environmental performance of products & businesses-substantiating claims" [16], companies will in the future be increasingly required to substantiate information on the environmental footprint of products or services using standardized quantification methods. The aim here is also to make environmental claims more reliable, comparable and verifiable throughout the EU and thus to reduce "greenwashing" and strengthen trust in environmentally relevant information. While DPP's overall contribution to facilitating circularity appears to be relatively clear and policy is currently moving the topic more into the spotlight, a widely applicable and holistic DPP-approach has not yet been established in practice. Accordingly, there are no finalized concepts at the political level as to how a DPP affects different stakeholders. However, there are some approaches and ideas on how the DPP could be implemented.

For instance, at the level of the EU's Member States, the German Government has picked up EU discussions on the DPP. According to the German Federal Ministry for the Environment, Nature Conservation and Nuclear Safety (BMU) [17], the digital product passport is defined as a data set that summarizes the components, materials and chemical substances or also information on repairability, spare parts or proper disposal for a product. The data originate from all phases of the product life cycle and are to be used for the optimization of design, production, use and disposal. The structuring of environmentally relevant data in a standardized, comparable format should enable all actors in the value and supply chain to work together towards a circular economy in a goal-oriented manner. At the same time, the digital product passport is intended as an important basis for more reliable consumer information and sustainable consumption decisions in both stationary and online retailing. According to the BMU, the DPP should in principle be applicable to all products and services as well as foodstuffs, with an initial focus on particularly resource- and energy-intensive goods [18]. These would include, for example, information and communications technology (ICT) products or products from other sectors with high energy and material consumption. Another study conducted by the European Policy Centre on behalf of the BMU that sketches possible ways of designing and implementing a DPP was published in late 2020. The aim of the study was to find "better coordination and exchange of information in value chains [to] enhance transparency while creating the basis for smart circular applications". The study suggests that the EU should start developing general guidelines for "tracking and mapping [. . .] products, materials and substances across value chains". A DPP should build on existing databases and information requirements and take into account the experience that companies have already gained in collecting information. The authors of the study propose the Commission to focus on textiles, electronics, construction, packaging, batteries and electric vehicles [5].

Due to the uncertain development of a DPP in the future and the lack of scientific debate on the DPP, this study seeks to work out design options of the DPP and important questions to be answered in the not-too-distant future regarding the implementation of the DPP. In so doing, we first show our step-wise approach (Section 2) and, then, describe the existing regime of regulated and voluntary product information tools focusing on the role of stakeholders (Sections 3.1 and 3.2). Intermediate results presented in Section 3.3 are examined in an actor-centered analysis on potential advantages gained through the DPP factoring in the most relevant stakeholder groups in a product's value chain. Lastly, in Section 4, we discuss our results with respect to the design of the DPP, and we focus on open questions, which need to be addressed in the not-too-distant future.

2. Materials and Methods

This study seeks to identify relevant points of discussion as regards the implementation of the DPP in order to maximize the socio-economic benefits across stakeholder groups. In so doing, we carried out a two-step approach, as shown in Figure 1.

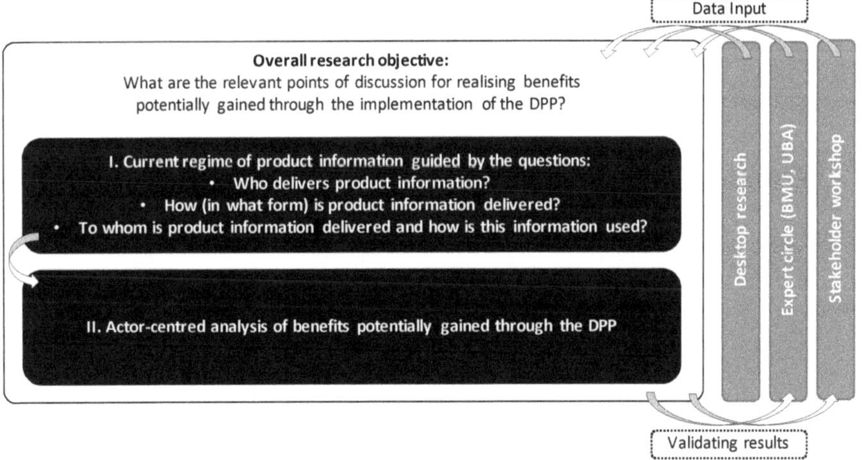

Figure 1. Research design.

First, we looked at the current regime of product information and factored in the following questions:

- Who delivers product information?

- How (in what form) is product information delivered?
- To whom is product information delivered and how is this information used?

By systematically reviewing relevant literature, we screened regulated and voluntary initiatives, which are implemented or developed from a variety of sectors in order to gain a rich overview of relevant factors to be taken into account when implementing the DPP as envisioned in the introduction of this article for the sake of providing more circular information. These findings on central characteristics of state-of-the-art information tools are then reflected in part two of the analysis: the actor-centered analysis. This part of the study will stimulate the discussion on the design of the DPP regarding the most relevant stakeholder types in a product's value chain: manufacturers, market surveillance, retailers, investors, repair shops, waste management companies.

Experts from the BMU and German Federal Environment Agency (UBA) were part of the project's expert circle and validated our findings periodically. In order to gain hands-on perspectives on the DPP, we also carried out a national expert stakeholder workshop in late 2020 as part of the project, this article is based on. More than 20 experts participated in the workshop, and the participants were selected in a way to cover a broad range of areas. This included experts from the BMU, the UBA and from the fields of standardization, digitalization, waste management, engineering and equipment manufacturing as well as academia. For the workshop, first project results were presented and discussed.

3. Results

Today, there are already a number of legal or voluntary information requirements in the area of product policy that determine information and information flows from point A to point B. At the EU level, information requirements exist for all phases of the product lifecycle, such as production, use, repair and disposal, but these requirements are mostly defined in a product-specific way. Results of the project, this article is based on Supplementary Materials.

3.1. Regulated Product Information

An illustrative example for current information flow regimes is the EU's energy labeling framework regulation, which defines a mandatory label and information obligations for selected product groups at the time of "placing on the market" (first time a product is made available on the EU market). With the status of March 2021, 15 product groups require an energy label [19]. Accordingly, product group or model-specific information must be published both on a label and on product data sheets. In the respective product group-specific implementation measures, the contents and information are further specified. For example, the label for refrigerators must include the manufacturer's name, the efficiency class, the electricity consumption per year, the volume of the refrigerator/freezer compartment and the maximum noise level for the corresponding model. The product data sheet, which must also be provided by the supplier, contains further information in addition to the information on the label, such as the exact design or duration of the manufacturer's guarantee. In addition, the Directive obliges suppliers to enter the information in the product data sheet and other data ("technical documentation") into an official digital EU database (EU Product Registration database for Energy Labelling, EPREL) via a special input page. This consists of both, a public part (for end users, among others) and a non-public part, which are only accessible to the European Commission and market surveillance authorities [20]. Apart from market surveillance, investors are a key target group of product information compiled by manufacturers. In particular, the Energy Label helps investors (including the public purse) to make conscious purchasing decisions [21], and the Label's recent revision of the scaling system is supposed to deliver higher efficiency gains through a more comprehensible labeling scheme. Retailers may also use the product information in sales talks, particularly those accompanied by the Energy Label. It should also be acknowledged that retailers do not enter or provide any new information, but they are responsible for ensuring that labels are placed on the respective products. To a very limited extent, repair companies and waste management companies can also benefit from the (limited) information by being able to verify certain aspects of the product.

Registration with the EPREL data is already mandatory as of February 2021 for the following product groups: air conditioners, household cooking appliances, household dishwashers, space heaters and water heaters, light bulbs, individual space heaters, household refrigeration appliances, commercial refrigeration appliances, solid fuel boilers, televisions, tumble dryers, residential ventilation appliances, and household washing machines [20]. In addition, since March 2021, consumers can also use the product database for the relevant public information on energy labels and product data sheets through a QR code that is printed on the label of some first product groups. Figure 2 below schematically illustrates the general structure of the EPREL product database.

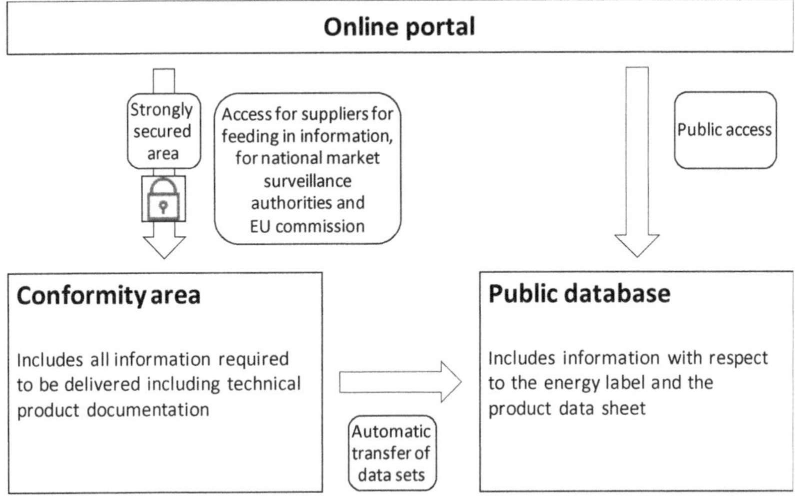

Figure 2. Schematic figure of the EPREL database (based on BMWi 2019 [22]).

The accessibility of the database for (potential) investors via an easy-to-use QR code is important to deliver information immediately at the point of sale, where a conventional website with cumbersome data entry would be of less help to investors.

In addition to the framework regulation on energy labeling, other EU regulations also contain subject-specific information and reporting obligations that differ more or less significantly depending on product and target group. For example, the EU Ecodesign Directive 2009/125/EC for energy-related products and appliances and its product group-specific implementing measures include, as does the closely linked EU framework regulation on energy labeling, information obligations at the time of "placing on the market". While aspects of circular economy and on repair options are increasingly included in the Ecodesign Directive, a central database has not been used for this purpose yet and a systematic data flow has not been prescribed. The information only has to be publicly available on a website of the manufacturer, importer or authorized representative. Another example for information requirements is the REACH Regulation EC 1907/2006 (REACH stands for Registration, Evaluation, Authorisation and Restriction of Chemicals). It includes safety data sheets for chemicals and further information on substances and mixtures and in particular on hazardous ingredients. Chemicals manufactured in the EU or imported into the internal market must be registered. The safety data sheets are primarily intended for persons who are in direct contact with the substances. This information must be provided either in electronic form or printed on paper and is intended to help protect health and the environment. In addition, the SCIP database ("database for information on Substances of Concern In articles as such or in complex objects (Products)") will be set up for SVHC ("substances of very high concern") in 2021 [23]. Suppliers will be required to provide their information to the European Chemicals Agency (ECHA). The aim of the database established is to provide operators of waste treatment plants with information on SVHCs in order to be able to separate them if necessary and to ensure high-quality recycling. Also focusing on chemicals, the Regulation EC/1272/2008 on classification, labelling and packaging (CLP) of substances and mixtures based on the United Nations Globally Harmonized System (GHS) has defined obligations for labeling [24]. Moreover, for this purpose, the European Chemicals Agency (ECHA) maintains a database on the classification and labeling of notified and registered substances. The Waste of Electrical and Electronic Equipment (WEEE) Directive 2012/19/EU is another example for information requirements. The Directive established obligations for electrical and electronic equipment, in particular with regard to the provision of information for recycling companies and operators of treatment facilities. This can be done by means of printed manuals or in electronic form. In addition, EU member states are required to establish a WEEE producer register. The EU Packaging and Packaging Waste Directive EU/2018/852 stipulates that clearly legible markings on materials in the packaging must be attached to the product; the Fertilizer Regulation EU/2019/1009 requires manufacturers to publish information on various product properties (storage conditions, volume, ingredients, etc.) on the product or in an accompanying document. The End-of-Life Vehicles Directive 2000/53/EC specifically regulates the publication of information on the dismantling, storage and testing of reused parts in end-of-life vehicles. In the international dismantling information system IDIS ("International Dismantling Information System"), vehicle manufacturers can deposit data to support disposal companies in the environmentally friendly treatment of end-of-life vehicles [25]. Another data collection system for vehicle manufacturers is the IMDS (International Material Data System), in which all materials used in the manufacture of a vehicle are collected [26]. In this context, the use of the IMDS should make it possible to fulfill the obligations imposed on the automotive industry by national and international standards, laws and regulations [27]. In addition to the examples presented, there are also various other approaches to data collection and presentation, such as the EU-wide standardized food labeling.

3.2. *Voluntary Product Information Initiatives*

In addition to the regulatory requirements, there are also numerous ideas and concepts on how (parts of) a digital product passport can be implemented. Some of these are already being implemented. One example is the concept of Material Passports. In more recent discussions, MPs have been developed with special focus on the building sector. Even though this concept is not necessarily restricted to construction materials only [28], buildings appear to be the central area of application so far. As part of the EU-funded research project focusing on reversing building design, partners develop an electronic Material Passport Platform as a one-stop-shop for material information provided by manufacturers and suppliers [29]. It is considered as record or documentation of properties of materials in order to facilitate recycling and reuse [30]. Hence, Material Passports increase transparency on the circularity characteristics of building materials and information includes, amongst others, data from technical data sheets or environmental product declarations (EPD). As soon as the a building is decommissioned, information can be made available to contracted deconstruction firms [31].

Technical documentation can be regulated, as for those product groups addressed under the Energy Labelling Directive. EPDs are generally voluntary and based on a life cycle assessment providing extensive quantitative and (third-party) verified information on environmental impacts without evaluating or judging them [32]. In Germany, EPDs have so far been used in practice also in particular for the comprehensive description of the environmental performance of building products. The environmental impacts of production, use and disposal are characterized according to internationally recognized conventions, resulting in key figures such as greenhouse potential in CO_2 equivalents, water consumption, waste production, ozone depletion potential or acidification potential [33]. In this way, EPDs should, for example, specifically facilitate the selection of materials in construction and form a basis for the documentation of the building materials used in the building (e.g., by means of a building passport) [34]. As regards the Material Passport Platform, the cross-referencing to other information tools shows that developers do not want to design new tools from scratch, but they also seek to build on existing information tools and embed this information for their purposes. Due to increased transparency, architects or builders can make use of materials with more circular characteristics.

Building information modeling (BIM) is seen as a vehicle to compile more comprehensive information on the entire building level (in contrast to the material level). BIM is a tool for networked planning, execution and management of buildings and may function also as an inventory database on the building level (in contrast to the component level). According to Honic et al. "the main results obtained from the BIM-supported MP is the total material composition of the building [...], which contrasts the share of recyclable materials with the share of waste created by the building" [35]. A challenge for MPs might be the feeding of material information continuously. For instance [36], state that steel used in buildings can, in general, be re-used without substantial testing in laboratories. However, if steel is exposed to fire, its characteristics may change, which is why the usage history of building materials can become important [36]. Such expositions but also major refurbishments, which could alter materials in buildings, would and could—ideally—be updated [31].

In addition to the MP, other concepts exist such as the cradle-to-cradle passport (C2C-passport). For example, the Danish shipping company Maersk already makes use of a C2C-passport for part of its own fleet of ships. The C-2-C-concept is based on a proprietary approach developed by McDonough Braungart Design Chemistry (MBDC). In 2010, MBDC transferred the certification program to the non-profit Cradle to Cradle Products Innovation Institute (C2CPII), which has since acted as a third-party certification body. The objective is to recycle materials used at the end of a product's life. Maersk's passport shows, for example, which materials are used in which location of a ship and provides, for instance, information about quality differences in the steel used. For Maersk, some of the key tasks were to develop a database for material information and to encourage suppliers to make complex material information (including its composition) available and feed it into the database. Materials should then be able to be located directly in a 3D model of the ship, which is why the passport already plays an important role in the development phase [37]. For ship owners or operators, this increases transparency and allows to identify potentials for reusing existing (and already purchased) materials. In the end, this may decrease material inputs and potentially overall costs for new ships, even though costs for training staff and deconstructing ships as well as testing steel characteristics will have to be added. As regards the C2C-passport, there is a direct (financial) interest in designing ships in a transparent way, which might be a different case for actors in the construction sector.

The comprehensive digitization of industrial production is known under the terminology of *Industrie 4.0*. In this context, the concept of the "asset administration shell" (AAS) was developed to systematically record and retrieve data on manufacturing equipment [38]. The AAS represents a digital image of the real production object, which is often also referred to as a "digital twin". The AAS, thus, opens up the conceptual link between the real and digital worlds. So far, this has been used primarily in progressive industrial companies and above all to optimize internal industrial production processes and procedures. Reference Architecture Model 4.0 (RAMI 4.0) is the (underlying) conceptual basis for data collection, which is based in principle on the Smart Grid Architecture Model (SGAM) established in the energy sector. In principle, the more relevant data is stored in the AAS, the more precise the mapping of the digital twin. Data (if available) can be mapped over the complete product life cycle, from development to the end of the product's life. Industry-internal information and communication technologies and IoT-technologies (Internet of Things) systems can thereby continuously capture and store data in real time so that the AAS can correspond with the real object as best as possible at any time. Data sets can, for example, consist of pre-configurations of production machines, material properties of intermediate products [39], limit values for use (e.g., maximum speed, highest possible operating temperature) or manuals, CAD drawings, key production figures (for example, target and actual values) or maintenance information [40,41]. However, the concepts of RAMI 4.0 and the AAS have so far been geared primarily toward use within highly networked *Industrie 4.0* areas. The AAS has therefore so far been used primarily in the production of complex production objects to create a network between appropriately equipped suppliers, integrators, machine manufacturers and other industrial users (cf. [38]). In theory, suppliers, integrators and manufacturers may benefit from increased information flows from the usage phase in order to improve product performance and for carrying out predictive maintenance.

3.3. *Key Takeaways from Regulated and Voluntary Product Information Initiatives*

All in all, a relatively clear picture emerges from the status quo analysis and from the different concepts and initiatives:

- Manufacturers and suppliers are, generally, the main actors to provide the specific product information. As regards other actors and, especially, retailers, they only forward relevant information but do not create new product data, which, in the end, means that the data flow in unidirectional.
- An exception of this is discussed for the Asset Administration Shell and also for the Material Passport, both of which, at least, discuss a more multidirectional information flow. In particular, the AAS is a good example that factors in current trends in digitalization (or IoT). Still, based on these findings, it can be assumed that acquiring product information *during* the use phase of a product is challenging in particular.
- Relevant information is supplied in a variety of forms including technical and safety data sheets (hard copies), labels and on the internet through websites or data portals.
- Online databases may contain confidential and non-confidential information, which can be accessible to selected user groups in a product value chain including manufacturers, market surveillance, retailers, investors, repair shops and waste management companies.
- Product information can be relevant to different user groups, but with different levels of detail, while market surveillance authorities need to have a clear overall picture with relatively detailed information, investors need simpler (and less detailed) information for their purchasing decisions.
- The development of business models (in delivering better product data regarding product circularity) is key to create acceptance, especially for manufacturers; should a manufacturer see a business case in product information (such as in the Maersk example), data compilation might be accompanied by an intrinsic motivation of the manufacturer.

These intermediate findings need to be taken into account as regards the potential benefits for each stakeholder group discussed in the next Section. As a summary, an overview of the different approaches compared to the currently discussed design of the DPP is illustrated in Table 1.

Table 1. Comparison of information tools.

	Digital Product Passport	Energy Label	Material Passport	C2C-Passport	AAS
Status	Pre-conceptual phase	Implemented	Demonstration	Implemented	Demonstration
Product Category	In theory, for all products discussed	Energy-related products	Building materials	e.g., Ships	Manufacturing equipment for *Industrie 4.0*

Table 1. *Cont.*

	Digital Product Passport	Energy Label	Material Passport	C2C-Passport	AAS
Key Information Categories	Origin, composition, repair and dismantling, handling at the end of its service life	Energy consumption, technical aspects	Information on reuse of materials; cross-ref. to data sheets, EPDs	Location of material use, material characteristics	Various
Life-Cycle Phase Targeted	Production, repair, disposal to complete life-cycle	Use phase	End-of-life	Production, repair, disposal	Complete life-cycle
Data Tool	unknown	EPREL	Materials Passport Platform, BIM discussed	Database for material information, 3D modelling	RAMI 4.0/AAS
Information Providers	Suppliers, manufacturers	Manufacturers	Manufacturers	Suppliers, manufacturer	Suppliers, manufacturers, (industrial) users
Target Groups	Market surveillance, consumers, repair shops, waste operators	Market surveillance, consumers	Architects, builders, deconstruction companies	For example, ship operator/owner	Suppliers, manufacturers, (industrial) users

3.4. Preliminary Actor-Centred Analysis of Potential Benefits Delivered by the DPP

Moreover, for the Digital Product Passport, manufacturers will likely remain the central suppliers of product information. Hence, additional (transaction) costs incurred due to further information demands (potentially also to be requested from suppliers) need to be kept at a minimum, even though it should be acknowledged that learning effects reduce the administrative costs in the longer run (cf. [42]), and trends in digitization (IoT, blockchain, machine learning) ease information gathering. Synergies should be seen with recent legislative developments, e.g., in Germany on the country's Supply Chain Act and similar initiatives on the EU level [43]. The DPP may help to provide a more consistent and untangled overall framework for manufacturers to deliver product information, but this would require a comprehensive integration of existing regulation and could be regarded as a challenging undertaking given that several of the above-mentioned regulations are administered by different Directorate Generals of the Commission. Still, gradually, the DPP may help to switch from mixed physical and digital information to a digital-only information supply including technical and safety data sheets. However, for this, it would also need to be ensured that target groups have the equipment necessary to really gain information access. In order to increase the motivation of manufacturers to deliver more circular information, attractive circular business models would need to be incentivized as well. This can also include that IoT-equipped products deliver information for manufacturers enabling them to expand their business model (e.g., predictive maintenance) as envisaged for the AAS. The Energy Label is also a success as it offers sustainable manufacturers to showcase their products' advantages in terms of sustainability and circularity and EPREL has high security standards which exacerbate data theft. Given that the DPP is supposed to be available for a variety of products, information requirements would need to be analyzed in a sector- or product-specific way (e.g., through a feasibility study) and manufacturers need to perceive a DPP infrastructure as a reliable and trustworthy system.

Market surveillance authorities can use product information to monitor whether manufacturers meet product standards in practice, also to protect manufacturers complying with standards against unfair competition. For such authorities, a central system, in which all information is organized, might be extremely helpful. In this respect, the EPREL database can be considered a good example as it is designed to contain selected regulated information. In our stakeholder workshop, experts argued that the digital product passport should also be seen as a part of a substance inventory, which takes stock of goods that are a" valuable secondary raw materials reservoir" and a "capital stock of the future" [44].

Retailers can use the improved information provided by a product passport to make their product range more customer-oriented and sustainable and to provide a corresponding range of information at the point of sale. Here, too, it plays a major role, which data retailers receive and to what extent this can be used in customer advice. In addition to retailers, contributors to the common good economy (second hand stores, etc.) should also be considered, as they can offer remanufactured products that are generally still usable. For them, the DPP may help if information from repair shops can be fed into the product documentation. Moreover, information on how a product has been used would also be largely beneficial as it would increase the trust of buyer in second-hand products. However, the question is what type of information can overcome barriers to purchasing second-hand products and how can the information be fed into the DPP. Amongst other, this may require the continuous multidirectional feeding of product-specific (in contrast to model-specific) information resembling the architecture of the AAS (which is largely envisioned for *Industrie 4.0*).

The key potential benefit of the DPP for product users is transparency, and private and institutional customers can make more conscious purchasing decisions. By differentiating between end-users, the role of green or sustainable public procurement should also be acknowledged as the public purse has a huge potential to transform products markets due to its buying power [45]. Products may reveal high social and ecological costs associated with production and customers are given the opportunity to buy products with a low socio-environmental footprint. Further valuable product information for customers may include the repairability and the end-of-life handling. However, it remains to be seen how information or data will be processed and made available to lay people. In order for customers to make sustainable purchasing decisions, information needs to be accessed with least possible effort. For instance, as regards the EU's Energy Label, the well-known scaling system (green to red arrows) visible to customers helps to easily differentiate between efficient and inefficient energy-related products, while disclosing only (standardized) energy consumption data (e.g., in kWh/a) would not be considered helpful by most users. An existing system for simple product identification for retail products, for example, is based on the "Global Trade Item Number" (GTIN), i.e., an identification number that can be used to uniquely identify many types of trade units. It must be mentioned here that this system has not yet been used for product-specific recording but rather for identification at the product group or model level. In any case, it is absolutely essential for a digital product passport that a product group, the model or, in perspective, even each individual product is clearly and easily identifiable. As with the EPREL database, for example, data could then be accessed directly via the individual item, e.g., via bar/QR codes or RFID tags on the product or product label (RFID stands for Radio Frequency Identification; small/tiny chips allow for wireless transfer of data). It would also make sense for consumers to be able to understand the information provided, including the language and meaning of the information, by making product features available via apps, websites or augmented reality, for example.

In contrast to product users, repair shops are dependent on precisely disaggregated information about repairs and spare parts, while information on socio-ecological effects associated with production is hardly a concern for them. Repair information is already required for some products (e.g., cars), and an extension could result in a rise of repair shops for many other products. An essential step will be that EU and national regulations require products to be manufactured in a way that factors in circularity and the right-to-repair. If in parallel, consumers are aware about the repairability of their products, this may strengthen the business model of repair shops.

In addition, companies from the waste management sector may also be interested in highly disaggregated data, which usually plays a minor role for consumers, for example. In particular, materials (and combinations) included in products, dismantling information and end-of-life handling will be of relevance. Through such information, dismantling costs can be reduced, and by selling recycled materials at higher qualities, revenues can be increased. If repair companies exchange certain components in a product, compositions of new materials used may also be relevant for waste companies.

4. Discussion

The Digital Product Passport seeks to facilitate a circular economy and a low carbon transition acknowledged by the EU [4] and the German BMU [17]. It is supposed to deliver information on the origin, composition, repair and dismantling options of a product, as well as on its handling at the end of its service life [3]. However, there are several open questions regarding the DPP's final design and its implementation. For instance, a long-time grown regime of diverse information requirements already exists, in which the DPP needs to be fitted into.

Having looked at certain parts of this existing landscape form a bird's eye perspective, we found that manufacturers are the most important source of product information. This means that any future DPP information requirements should be ideally designed in a way that manufacturers and other stakeholders perceive them as an advantage and not as an additional burden, in order to create business models and intrinsic motivation. If additional information obligations are imposed, they should create as much as possible synergies with other compliance regulation (cf. [43]). Therefore, the initial DPP approach should build-up on existing systems of regulations [5] also acknowledging technology trends as well as learning effects for information supply [42].

For instance, under the Waste Framework Directive, companies supplying products containing SVHC (above certain concentrations) already must supply selected information on these articles to a database made available to waste operators and consumers [23]. Under the Energy Labelling Directive, manufacturers of refrigerators have to supply a variety of information (e.g., efficiency class, electricity consumption per year, the maximum noise level for the corresponding model). However, this information mostly focuses on the use phase of a product and have to be fed into the EPREL database. In contrast to that, the Waste of Electrical and Electronic Equipment goes beyond the Energy Labelling Directive's product scope and mandates manufacturers to deliver information on equipment disposal and handling at the end of its life, while the End-of-Life Vehicles Directives focuses on similar information types (e.g., dismantling information) but for a particular product group. For the DPP, a key question will be how to organize an optimized and synergetic data flow with the existing framework of regulatory efforts for manufacturers, which really are the core stakeholder group, at the moment. In contrast to the regulated information flows, there are also voluntary initiatives on the market or in development. In our study, we selected some information tools, which seek to contribute to a circular economy. They also differ from each other. Similar to EPREL or SVHC, they make use of a digital system to compile, feed in and retrieve data or information.

It might be helpful to investigate further on the existing information tools in order to find out what information are technically feasible to be supplied for a DPP. An option to reduce the administrative burden of manufacturers can be to, initially, develop an approach that integrates existing information requirements in a smart way, where a single point of information brings together all existing information with high security standards and provides them according to different access rights to specific stakeholder groups (cf. [7]).

Thus, this single point of information will be fed by manufacturers with minimal transaction costs for changing information supply. In other words, the information requirements (mandated in various regulations and directives) remain the same, so there is not additional effort to compile new information for manufactures, only the point to enter the relevant information might differ.

As regards the basic technical infrastructure necessary to implement the DPP, the experiences from the EPREL database as well as the Asset Administration Shell deserve some more attention. One of the basic key features of EPREL is that confidential information and non-confidential information can be fed into the database, which is relevant if information may be mandatory from the perspective of a market surveillance agency but not for other stakeholders; some information might also have to be shielded from competitors (e.g., extraction/production location of certain inputs for consumer goods). The AAS, considered largely for advancing *Industrie 4.0* and addressing respective equipment, could even provide a basis for multidirectional data exchange regarding single products. This would be interesting e.g., for the information exchange between repair shops and waste operators, especially if the concept of the AAS could be transferred and adjusted for non-industrial purposes. For instance, if particular spare parts are used differing from the original product set up, waste operators could require adjusted product information for recycling purposes. Other opportunities for the multidirectional information flow might also exist and, thus, information feedbacks between different stakeholders should be explored factoring in, e.g., advances in the field of digitization, in general, and IoT, in particular.

All in all, how to generate data during the use phase will remain extraordinarily challenging. It would give not only investors the opportunity to exchange components in advance before more serious damage occur, but is also offers equipment providers to extent business models for instance through predictive maintenance and receiving data in order to improve technology. At present, in most traditional sectors where a manufacturer "just sells" a product to an investor, there is hardly any business case for the manufacturer further down in their product's value chain. So conventional and linear business models still dominate in most sectors. However, the example of the company Maersk suggests that the company hopes to identify corporate sustainability information and new revenue streams or reduce costs at the same time through being better able to identify certain products in ships built. Likewise, the Material Passport factoring in Building Information Modelling may help to break the existing paradigm in construction works helping to generate information during the use phase.

Apart from questions around the existing (regulatory and also voluntary) information landscape and the technical infrastructure, an essential aspect is to focus also on the question how to increase general attractiveness of the DPP to users/investors. For instance, the Energy Label also enjoys broad stakeholder support as it offers manufacturers to illustrate the uniqueness and benefits of their certain product's characteristics to investors (apart from energy efficiency, also noise pollution). However, how can the DPP create similar transparency as regards the circularity of products in order to contribute to the objectives of the European Consumer Agenda [15]? In other words: How will customers know and easily understand which refrigerator belongs to the most "circular" or sustainable ones? Product information only available to stakeholders further down the value chain (repair shops, waste operators) is important for a circular economy but not necessarily to persuade investors to invest in a certain product. Hence, in order to make sustainable choices, consumers need transparent, simple information. If the DPP seeks to raise the awareness of a product's circularity characteristics, the EU needs to find out how this can be achieved (again, without increasing the administrative burden, in parallel).

With the discussion on a digital product passport gaining momentum, there is currently an ideal window of opportunity to bundle ideas at the European level and derive initial options for action as well as further research approaches [3,4]. Scientific feasibility studies should be carried out as soon as possible on how to implement a digital product passport [5]. The German Environmental Agency began to initiate such a study on textiles and energy-related products, but further research will also have to scrutinize EU-wide conditions in various pilot projects. An analysis of the data needs of various stakeholder groups is essential but also whether these wishes can be realistically met and how taking into account different manufacturers. Since the concept of a digital product passport is still relatively new, there are currently several aspects to be clarified promptly by additional research activities for rapid and concrete implementation. These include, for example, the more precise selection of product groups to be prioritized and thus the question of which products are particularly suitable for the fastest possible introduction of a product passport system. The assessment of various experts and interest groups also still differs greatly in some cases on the question which criteria and precise data requirements should be addressed by a digital product passport. Therefore, a detailed stakeholder analysis including a differentiation, at least, regarding certain subtypes (e.g., SME vs. large companies) should be conducted and is also necessary at the beginning of further research activities in order to determine the respective information needs and acceptance factors more precisely.

In order to involve the relevant stakeholders in this process and promote acceptance, an early exchange within the framework of a scientifically accompanied consultation process is therefore recommended so that opportunities, interests, obstacles and challenges can be identified through active participation. Stating the obvious, the DPP will not be a silver bullet for achieving a circular economy alone, but its realization might make particular sense to form a key instrument in a well-orchestrated policy mix [12].

5. Conclusions

In order to identify the relevant points of discussion regarding the implementation of the Digital Product Passport, we first screened the current landscape of existing information tools. From the tools scrutinized we were able to draw some key lessons:

- Manufacturers and suppliers are, generally, the main actors to provide the specific product information;
- The Asset Administration Shell and the Material Passport are interesting use cases and examples for the management of multidirectional information flows;
- Relevant and comprehensive product information is supplied already today but in a variety of formats;
- Online databases with dedicated access control may contain and handle confidential and non-confidential information;
- Product information can be relevant to different user groups but with different levels of detail;
- The development of business models (in delivering better product data regarding product circularity) is key to create acceptance, especially for manufacturers.

In a second step, these lessons were fed into our actor-centered analysis helping to carve out achievable benefits by the DPP, which depend on the overall implementation design of the instrument. In the previous chapter, we discussed that the DPP may be integrated into existing systems of information regulations but that it will be relevant to organize synergetic data flows with the existing framework. In so doing, a single point of information could bring together all existing information with high security standards and provide them according to different access rights to specific stakeholder groups. This single point of information could be fed by manufacturers with minimal transaction costs for changing information supply. Apart from that, the multidirectional information flow is highly interesting as this, e.g., would enable the information exchange between repair shops and waste operators. However, the collection of data during the use phase will remain extraordinarily challenging, though probably more relevant and feasible for some products (e.g., high value products with longer product lifetime) compared to others. Besides, the role of investors must be factored in, and the DPP should ideally help investors to better understand which products belong to the most sustainable ones in their respective product group. However, as described, all those potential design options still need further scientific investigation concerning their suitability for real-life use.

Considering all gained perspectives and results, the DPP is a very promising policy instrument that is correspondingly linked with high expectations by many stakeholders. However, being at an early stage of the discussion, several open issues need to be addressed before a Digital Product Passport can be implemented on a large scale. With this paper, we hope to initiate a broader scientific discussion and that further research take on these challenging questions to provide orientation for the DPP's design. If implemented carefully in a sense that visibly increases the benefits for different actor types and ideally also reduces costs or efforts, there is a strong potential to drive sustainable product policy in a more circular direction. Closing the material loop in the sense of a more holistic ecodesign can mean that the EU's demand for new raw materials can be reduced while increasing independence of the EU from less trustworthy suppliers at the same time (also increasing leverage in other policy fields). Information on better product usage and repair may result in innovative new circular business models in the EU extending the lifetime of products and creating also new efficiency and job opportunities. Within the EU market, the DPP in combination with complementary regulation may help innovative manufactures to stand out from competitors that hardly care about circularity. At the same time, given the EU market's strong international role in and influence on manufacturing worldwide, the DPP (in combination with other instruments, such as ecodesign) may also function as a further starting signal to transform production systems globally towards more sustainability. In this context, the DPP could be seen also as part of a complex puzzle to lower the divide between more industrialized and less industrialized countries in the sense of the SDGs.

Supplementary Materials: Results of the project this article is based on are available in Götz, T.; Adisorn, T.; Tholen, L. Der Digitale Produktpass als Politik-Konzept: Kurzstudie im Rahmen der Umweltpolitischen Digitalagenda des Bundesministeriums für Umwelt, Naturschutz und nukleare Sicherheit (BMU); Wuppertal Report; Wuppertal Institut für Klima, Umwelt, Energie: Wuppertal, Germany, 2021; Volume 20, p. 44, also available online at https://wupperinst.org/a/wi/a/s/ad/7315 (accessed on 16 April 2021).

Author Contributions: Conceptualization, T.A. and T.G.; methodology, T.A., L.T. and T.G.; formal analysis, T.A. and L.T.; investigation, T.A., L.T. and T.G.; writing—original draft preparation, T.A., L.T. and T.G.; writing—review and editing, T.A., L.T. and T.G.; visualization, T.A.; supervision, T.G., T.A.; project administration, T.G. and T.A. All authors have read and agreed to the published version of the manuscript.

Funding: This article is based on the project "Environmental Policy and Digitalization", which was funded by the German Federal Ministry for the Environment, Nature Conservation and Nuclear Safety. The responsibility for the content of this publication lies with the authors.

Institutional Review Board Statement: Not applicable.

Informed Consent Statement: Not applicable.

Data Availability Statement: Not applicable.

Acknowledgments: We would like to thank the various experts from the German Federal Ministry for the Environment, Nature Conservation and Nuclear Safety as well as the German Federal Environment Agency and participants in the stakeholder workshop for their valuable input contributing to the results of the overall project, this article is based on.

Conflicts of Interest: The authors declare no conflict of interest.

Abbreviations

AAS	Asset Administration Shell
BIM	Building Information Modeling
BMU	German Federal Ministry for the Environment, Nature Conservation and Nuclear Safety
C2C	Cradle to Cradle
CLP	Regulation on Classification, Labelling and Packaging
DPP	Digital Product Passport

ECHA	European Chemicals Agency
EPD	Environmental Product Declaration
EPREL	EU Product Registration database for Energy Labelling
GLS	Globally Harmonized System
GTIN	Global Trade Item Number
ICT	information and communication technology
IDIS	International Dismantling Information System
IMDS	International Material Data System
IoT	Internet of Things
RAMI	Reference Architecture Model
REACH	Regulation concerning Registration, Evaluation, Authorisation and Restriction of Chemicals
RFID	Radio Frequency Identification
SCIP	Substances of Concern In articles as such or in complex objects (Products)
SDG	Sustainable Development Goals
SGAM	Smart Grid Architecture Model
SPI	Sustainable Product Initiative
SVHC	Substances of very high concern
WEEE	Waste of Electrical and Electronic Equipment Directive

References

1. United Nations. *Transforming Our World: The 2030 Agenda for Sustainable Development*; United Nations: New York, NY, USA, 2015.
2. United Nations, Department of Economic and Social Affairs. The 17 Goals. Available online: https://sdgs.un.org/goals (accessed on 20 March 2021).
3. European Commission. *The European Green Deal. COM(2019) 640 Final*; European Commission: Brussels, Belgium, 2019.
4. European Commission. *Circular Eonomy Action Plan. For a Cleaner and More Competitive Europe*; European Commission: Brussels, Belgium, 2020.
5. Hedberg, A.; Šipka, S. *Towards a Green, Competitive and Resilient EU Economy: How Can Digitalisation Help?* European Policy Centre: Brussels, Belgium, 2020; p. 22.
6. Council of the European Union. *Draft Council Conclusions on Making the Recovery Circular and Green*; Council of the European Union: Brussels, Belgium, 2020.
7. Götz, T.; Adisorn, T.; Tholen, L. *Der Digitale Produktpass als Politik-Konzept: Kurzstudie im Rahmen der Umweltpolitischen Digitalagenda des Bundesministeriums für Umwelt, Naturschutz und nukleare Sicherheit (BMU)*; Wuppertal Report; Wuppertal Institut für Klima, Umwelt, Energie: Wuppertal, Germany, 2021; Volume 20, p. 44.
8. Gligoric, N.; Krco, S.; Hakola, L.; Vehmas, K.; De, S.; Moessner, K.; Jansson, K.; Polenz, I.; van Kranenburg, R. SmartTags: IoT Product Passport for Circular Economy Based on Printed Sensors and Unique Item-Level Identifiers. *Sensors* **2019**, *19*, 586. [CrossRef] [PubMed]
9. Donetskaya, J.V.; Gatchin, Y.A. Development of Requirements for The Content of a Digital Passport and Design Solutions. *J. Phys. Conf. Ser.* **2021**, *1828*, 012102. [CrossRef]
10. Longo, S.; Cellura, M.; Cusenza, M.A.; Guarino, F.; Mistretta, M.; Panno, D.; D'Urso, C.; Leonardi, S.G.; Briguglio, N.; Tumminia, G.; et al. Life Cycle Assessment for Supporting Eco-Design: The Case Study of Sodium–Nickel Chloride Cells. *Energies* **2021**, *14*, 1897. [CrossRef]
11. Wielgosiński, G.; Czerwińska, J.; Szufa, S. Municipal Solid Waste Mass Balance as a Tool for Calculation of the Possibility of Implementing the Circular Economy Concept. *Energies* **2021**, *14*, 1811. [CrossRef]
12. Milios, L. Advancing to a Circular Economy: Three Essential Ingredients for a Comprehensive Policy Mix. *Sustain. Sci.* **2018**, *13*, 861–878. [CrossRef] [PubMed]
13. European Commission. Sustainable Product Initiative. Available online: https://ec.europa.eu/info/law/better-regulation/have-your-say/initiatives/12567-Sustainable-Products-Initiative (accessed on 2 March 2021).
14. European Commission. Consumer Policy—Strengthening the Role of Consumers in the Green Transition. Available online: https://ec.europa.eu/info/law/better-regulation/have-your-say/initiatives/12467-Empowering-the-consumer-for-the-green-transition (accessed on 20 March 2021).
15. European Commission. *New Consumer Agenda. Strengthening Ocnsumer Resilience for Sustainable Recovery. COM(20230)696 Final*; European Commission: Brussels, Belgium, 2020.
16. European Commission. Environmental Performance of Products & Businesses—Substantiating Claims. Available online: https://ec.europa.eu/info/law/better-regulation/have-your-say/initiatives/12511-Environmental-claims-based-on-environmental-footprint-methods (accessed on 20 March 2021).
17. Federal Ministry for the Environment, Nature Conservation, and Nuclear Safety. BMU Digital Policy Agenda for the Environment: Digital Product Passport. Available online: https://www.bmu.de/en/service/haeufige-fragen-faq/details-cluster/bmu-digital-policy-agenda-for-the-environment-digital-product-passport/ (accessed on 20 March 2021).
18. Federal Ministry for the Environment, Nature Conservation, and Nuclear Safety. Häufige Fragen (FAQ) Umweltpolitische Digitalagenda. Available online: https://www.bmu.de/faqs/umweltpolitische-digitalagenda-digitaler-produktpass (accessed on 20 March 2021).

19. European Commission. Rules and Requirements for Energy Labelling and Ecodesign. Available online: https://ec.europa.eu/info/energy-climate-change-environment/standards-tools-and-labels/products-labelling-rules-and-requirements/energy-label-and-ecodesign/rules-and-requirements_en (accessed on 19 March 2021).
20. European Commission. Product Database. Available online: https://ec.europa.eu/info/energy-climate-change-environment/standards-tools-and-labels/products-labelling-rules-and-requirements/energy-label-and-ecodesign/product-database_en (accessed on 19 March 2021).
21. Ecofys; Waide Strategic Efficiency; Öko-Institut; University of Coimbra; SEVEn; SoWatt. *Evaluation of the Energy Labelling Directive and Specific Aspects of the Ecodesign Directive*; Ecofys: Utrecht, The Netherlands, 2014.
22. Bundesministerium für Wirtschaft und Energie. *Die neue EU-Produktdatenbank EPREL. Ein Informationspapier für Hersteller und Importeure von Energieverbrauchsrelevanten Produkten*; Bundesministerium für Wirtschaft und Energie: Berlin, Germany, 2019; p. 7.
23. European Chemicals Agency. ECHA SCIP. Available online: https://echa.europa.eu/de/scip (accessed on 20 March 2021).
24. UNECE. About the GHS. Globally Harmonized System of Classification and Labelling of Chemicals (GHS). Available online: https://unece.org/about-ghs (accessed on 20 March 2021).
25. IDIS. International Dismantling Information System. Available online: https://www.idis2.com/ (accessed on 20 February 2021).
26. Material Data System IMDS Information Pages. Available online: https://public.mdsystem.com/en/web/imds-public-pages (accessed on 20 March 2021).
27. EntServ Deutschland IMDS Informationsseiten. Available online: https://www.mdsystem.com/imdsnt/startpage/index.jsp (accessed on 26 March 2021).
28. GXN Innovation; 3XN Architects; MT Højgaard; VIA Byggeri; Kingo Karlsen; Vugge til Vugge Danmark; henrik•innovation. *Building a Circular Future*; Danish Environmental Protection Agency: Copenhagen, Denmark, 2018.
29. Buildings as Material Banks (BAMB) Materials Passports. Available online: https://www.bamb2020.eu/topics/materials-passports/ (accessed on 11 March 2021).
30. Lescuere, L.M. Materials Passports: Optimising Value Recovery from Materials. *Proc. Inst. Civ. Eng. Waste Resour. Manag.* **2016**, *170*, 25–28. [CrossRef]
31. Heinrich, M.; Lang, W. *Material Passports—Best Practice*; Technical University of Munich: Munich, Germany, 2019.
32. Bundesministerium für Umwelt, Naturschutz und nukleare Sicherheit; Bundesverband der Deutschen Industrie e.V.; Umweltbundesamt. *Umweltinformationen Für Produkte Und Dienstleistungen Anforderungen—Instrumente—Beispiele*; Bundesministerium für Umwelt, Naturschutz und nukleare Sicherheit: Berlin, Germany, 2019.
33. Institut Bauen und Umwelt Veröffentlichte EPDs. Available online: https://ibu-epd.com/veroeffentlichte-epds/ (accessed on 15 March 2021).
34. Umweltbundesamt Umweltdeklaration von Bauprodukten. Available online: https://www.umweltbundesamt.de/themen/wirtschaft-konsum/produkte/bauprodukte/umweltdeklaration-von-bauprodukten (accessed on 15 March 2021).
35. Honic, M.; Kovacic, I.; Rechberger, H. Concept for a BIM-Based Material Passport for Buildings. *IOP Conf. Ser. Earth Environ. Sci.* **2019**, *225*. [CrossRef]
36. Smeets, A.; Wang, K.; Drewniok, M.P. Can Material Passports Lower Financial Barriers for Structural Steel Re-Use? *IOP Conf. Ser. Earth Environ. Sci.* **2019**, *225*, 012006. [CrossRef]
37. Sterling, J. Cradle to Cradle Passport. Available online: https://www.oecd.org/sti/ind/48354596.pdf (accessed on 8 April 2021).
38. Zentralverband Elektrotechnik- und Elektroindustrie (ZVEI). *Beispiele Zur Verwaltungsschale Der Industrie 4.0-Komponente - Basisteil. Fortentwicklung Des Referenzmodells Für Die Industrie 4.0-Komponente*; Zentralverband Elektrotechnik- und Elektroindustrie (ZVEI): Frankfurt am Main, Germany, 2016.
39. Müller, J.; Both, M. Hersteller Definieren Industrie-4.0-Verwaltungsschale Für Pumpen. Available online: https://www.chemietechnik.de/anlagentechnik/schuettguttechnik/pumpenhersteller-definieren-industrie-4-0-verwaltungsschale-fuer-fluessigkeits-und-vakuumpumpen.html (accessed on 3 March 2021).
40. Arnold, R.; Liebe, A. *Digitale Wertschöpfungsnetzwerke Und RAMI 4.0 Im Mittelstand*; Hessisches Ministerium für Wirtschaft, Energie und Landesentwicklung: Wiesbaden, Germany, 2018.
41. Kaspar, B. *Industrie 4.0: Technologieentwicklung Und Sicherheitstechniche Bewertung von Anwendungsszenarien.*; Bundesanstalt für Arbeitsschutz und Arbeitsmedizin: Dortmund/Berlin/Dresden, Germany, 2019.
42. Jantzen, J. *Environmental Expenditures in EU Industries. TIme Series Data for the Costs of Environmental Legilsation for Selected Industries over Time.*; European Commission: Brussels, Belgium, 2015.
43. Institut der deutschen Wirtschaft Braucht Deutschland ein strengeres Lieferkettengesetz? Available online: https://www.iwd.de/artikel/braucht-deutschland-ein-strengeres-lieferkettengesetz-505513/ (accessed on 31 March 2021).
44. Leibniz-Institut für ökologische Raumentwicklung; Wuppertal Institut; INTECUS. *Kartierung Des Anthropogenen Lagers in Deutschland Zur Optimierung Der Sekundärstoffwirtschaft*; UBA Texte 83/2015; Wuppertal Institut: Dessau-Roßlau, Germany, 2015.
45. European Commission Green Public Procurement. Available online: https://ec.europa.eu/environment/gpp/index_en.htm (accessed on 20 March 2021).

 energies

Article

Technological Solutions and Tools for Circular Bioeconomy in Low-Carbon Transition: Simulation Modeling of Rice Husks Gasification for CHP by Aspen PLUS V9 and Feasibility Study by Aspen Process Economic Analyzer

Diamantis Almpantis and Anastasia Zabaniotou *

Circular Bioeconomy and Sustainability Research Group, Department of Chemical Engineering, Engineering School, Aristotle University of Thessaloniki, 54 124 Thessaloniki, Greece; diamosalmpantis@gmail.com
* Correspondence: azampani@auth.gr; Tel.: +30-6945990604

Abstract: This study explored the suitability of simulation tools for accurately predicting fluidized bed gasification in various scenarios without disturbing the operational system, and dedicating time to experimentation, in the aim of benefiting the decision makers and investors of the low-carbon waste-based bioenergy sector, in accelerating circular bioeconomy solutions. More specifically, this study aimed to offer a customized circular bioeconomy solution for a rice processing residue. The objectives were the simulation and economic assessment of an air atmospheric fluidized bed gasification system fueled with rice husk, for combined heat and power generation, by using the tools of Aspen Plus V9, and the Aspen Process Economic Analyzer. The simulation model was based on the Gibbs energy minimization concept. The technological configurations of the SMARt-CHP technology were used. A parametric study was conducted to understand the influence of process variables on product yield, while three different scenarios were compared: (1) air gasification; (2) steam gasification; and (3) oxygen-steam gasification-based scenario. Simulated results show good accuracy for the prediction of H2 in syngas from air gasification, but not for the other gas components, especially regarding CO and CH4 content. It seems that the RGIBBS and Gibbs free minimization concept is far from simulating the operation of a fluidized bed gasifier. The air gasification scenario for a capacity of 25.000 t/y rice husk was assessed for its economic viability. The economic assessment resulted in net annual earnings of EUR 5.1 million and a positive annual revenue of EUR 168/(t/y), an excellent pay out time (POT = 0.21) and return of investment (ROI = 2.8). The results are dependent on the choices and assumptions made.

Keywords: rice husk; gasification; CHP; Aspen Plus; simulation; economic assessment; circular economy; low-carbon energy; waste-based bioenergy

Citation: Almpantis, D.; Zabaniotou, A. Technological Solutions and Tools for Circular Bioeconomy in Low-Carbon Transition: Simulation Modeling of Rice Husks Gasification for CHP by Aspen PLUS V9 and Feasibility Study by Aspen Process Economic Analyzer. *Energies* **2021**, *14*, 2006. https://doi.org/10.3390/en14072006

Academic Editors: Anna Mazzi and Jingzheng Ren

Received: 7 March 2021
Accepted: 26 March 2021
Published: 5 April 2021

Publisher's Note: MDPI stays neutral with regard to jurisdictional claims in published maps and institutional affiliations.

Copyright: © 2021 by the authors. Licensee MDPI, Basel, Switzerland. This article is an open access article distributed under the terms and conditions of the Creative Commons Attribution (CC BY) license (https://creativecommons.org/licenses/by/4.0/).

1. Introduction

The biocapacity of earth in biomass resources amounts to 172 billion t of dry matter that contains ten times more energy than the energy consumed worldwide [1] (Eurotex, 2020). This huge energy potential remains largely unexploited, as only 1/7 of the world's energy consumption is covered by biomass, mainly for traditional uses (combustion). However, 1 t of biomass is equivalent to about 0.4 t of fuel oil, only 3% of global energy needs are met by using available biomass [2].

Residues and waste from agricultural and industrial processes in Mediterranean countries, such as olive kernels or rice husks from agro-industrial plants, wine from wineries or fruit stones from fruit processing industries, are insufficiently used, resulting in a significant amount of waste left in the fields. Taking Greece as a Mediterranean case, although its total available biomass reaches approximately 7,500,000 t of crop residues (cereals, maize, cotton, tobacco, sunflower, twigs, vines), and 2,700,000 t of forest residues (branches, bark), in addition to significant amounts of residual biomass from energy crops, the largest percentage of this biomass remains unused, often causing many hazards (fires, spreading diseases) [2].

The European Union (EU) aims to increase biomass uses towards helping to achieve goals of renewable energy and greenhouse gas (GHG) emissions reduction. Facilities that use locally renewable energy sources, designed to supply local energy communities through micro- and small-scale units, are at the forefront of the EU energy strategy, while combined heat and power (CHP) production from agricultural waste and residue could be a viable way for the development of renewable, reliable, and affordable electricity, while improving waste management, contributing to sustainable agriculture, and implementing circular economy innovations [3].

The end-users of bioenergy can vary in scale, from households, school, public buildings and tourist complexes to district heating, and heat and steam production in agro-industrial facilities. Thus far, CHP biomass systems have been proven to be viable only at large scales that are supported by tariffs or green certificates. However, it is important to mention that large scale bioenergy demand for the scale-up of biomass availability may have some consequences on the environmental impacts that the bioenergy sector can create, due to the direct relationship between the biomass demand scale and the GHG profile of its production.

Sustainable small-scale biomass plants, which produce CHP, appear to be among the most promising techniques for decentralized energy production if they can operate sustainably. The small-scale units benefit from a flexible integrated technology system, with the possibility of the successful penetration into the electricity market, in the market, and the promotion of regional development and the strengthening of the agricultural sector. However, investments or long repayment times create obstacles to their implementation [3].

Gasification is the thermochemical conversion of biomass into gas-fuel through heat with a gasifier agent such as air, oxygen, or steam. Air is the most used gasification agent because it is cheap and readily available [4]. The syngas produced can be stabilized in quality, so it is easier to be used and has multiple uses compared to the original biomass from which it derives, in gas engines and gas turbines, or even as a power supplier for liquid fuel production [5]. Installed gasification units operating in different parts of the world are differentiated by the type of gasifier. Gasifiers fueled with organic materials and residual biomass may need to be specially designed for higher efficiencies, better economies, and a safe environment [5]. Although the smallest size of biomass particles is favorable, it is essential to consider that energy consumption to reduce particle size should reduce overall energy efficiency, therefore different types of gasifiers should be designed to handle different sizes of biomass particles [4].

Scope and Objectives of the Study

This study aimed to present a customized circular waste-based bioeconomy solution for a rice processing industrial sector, which is of great technological and commercial interest in many countries, and to support the use of simulation tools for the planning phase of bioenergy solutions within a circular bioeconomy. These tools are the Aspen PLUS V9 and Aspen Process Economic Analyzer for process simulation and economic assessment, respectively, which were used in this study for the air gasification-based CHP system fueled by rice residue.

The scientific objectives were: (1) the simulation of an atmospheric fluidized bed gasification (FBG) system fueled with rice husk (RH) for CHP generation with an Aspen Plus V9 simulation modeling (SM) tool; (2) the simulation of steam and (steam + oxygen) FBG scenarios to compare with the air FBG main scenario by using suitable indicators; and (3) the feasibility study of an air gasification-based unit with a capacity of 25,000 t/y RH using the Aspen Process Economic Analyzer.

This study does not intend to bring technical innovation beyond the state of the art on gasification and CHP technology. It is based on the SMARt-CHP innovative technology, a prototype of an FBG-based system designed and developed at Aristotle University, Greece, and funded by a European Commission LIFE+ project some years ago. The experimental proof of concept of RH gasification results are published elsewhere [6]. After collecting the experimental results and designing an FBG system, it was considered a useful move for the bioenergy sector to introduce simulation modeling (SM) to allow developers and users to examine the system operation, using different possible scenarios and conditions, and using less time-consuming tools for planning at higher technological readiness levels (TRLs).

2. Methodology

The simulation study was based on experimental data obtained at our laboratory by previous researchers [6]. The technology used was the SMARt CHP technology developed by our team and described in a previously published work [7].

The modeled flow diagram of the bioenergy system was developed by using the Aspen Plus software, which proposes appropriate devices for the process simulation at the proposed operating conditions.

A sensitivity analysis was performed to explore the relation of syngas product using indicators such as the equivalence ratio (ER), low heating value (LHV), cold efficiency (CCE) and cold gas efficiency (CGE) and the steam-to-biomass ratio (SBR) by selecting the gasification agent as the design variable. The comparison of the gasification efficiency in relation to the use of other gasifying agents (air, steam, and combination of oxygen-steam) was also performed.

An economic assessment was performed by using the Aspen Process Economic Analyzer software, estimating the economic indicators of fixed investment, total investment, annual operating costs and the net profit of the unit, as well as the return of investment (ROI) and pay out time (POT) indicators, in order to assess economic viability.

Finally, conclusions of the study were drawn, while assumptions and approaches considered in the calculations were commented on.

3. Materials and Methods

RH is of a huge reserve and availability at a low price in Greece. It is the by-product of the industrial processing of rice. It accounts for approximately 20 wt.% of bulk grain weight and is very often used as an alternative source of silica in ceramics [8]. It contains 70–80% organic substances such as cellulose, lignin, and 20–30% components such as silica, alkalis, and trace elements [9]. Due to its high calorific value, it can be used as fuel for energy production by gasification [10].

RH has a low inherent moisture content (<10 wt.%) and a C/N ratio >150, thus it is an appropriate fuel for thermochemical processing, such as gasification. Gasification generates the rice husk ash (RHA), which accounts for about 25% of the initial husk weight and causes environmental disposal problems [11].

3.1. Choice of Materials

RH used in this study was provided by Greek company "Agrino", which is the largest rice producer in Greece (5 t/h) (http://www.agroenergy.gr/content, accessed on 30 May 2020). This production accounts for RH production of approximately 20 wt.% of the total paddy weight (whole grain). Therefore, 5 t/h of paddy grain produces about 1 t/h (20%) of RH, and when it is gasified to generate energy, it generates also about 250 kg/h (25 wt.%) of ash, a volume containing around 45 kg (85–95%) of amorphous silica [12]. The ultimate, proximate, and chemical analysis of "Agrino" RH is presented in Table 1.

Table 1. Ultimate and proximate and chemical analysis of rice husk (RH) [6].

Ultimate Analysis (wt.%, dry)	
Carbon, C	36
Hydrogen, H	5.20
Oxygen, O	38.0
Nitrogen, N	0.3
Sulfur, S	0.04
Proximate Analysis (wt%, a.r)	
Moisture	n.a
Volatiles	67
Fixed carbon	15
Ash	18
Higher heated value (HHV) (MJ/kg)	14.50
Chemical Analysis (mg/kg, dry)	
Aluminum (Al)	115
Potassium (K)	2794
Sodium (Na)	72
Calcium (Ca)	1256
Silicon (Si)	81
Chlorine (Cl)	684
Magnesium (Mg)	383
Iron (Fe)	186
Phosphorus (P)	376
Titanium (Ti)	5
Manganese (Mn)	220

3.2. Choice of the Technology

The technological system used was the SMARt-CHP system that produces renewable CHP from waste-biomass and is used for waste management [7]. The electricity generated is either used on-site or it is supplied to the grid. The heat generated by the process is used to heat the industry's buildings.

SMARt-CHP is a technological output that is suitable for circular economy applications. It consists of a pilot fluidized bed gasifier coupled with an internal combustion engine (ICE). It was designed and developed in our laboratory, funded by an EU LIFE+ project (www.smartchp.eng.auth.gr, accessed on 30 May 2020). The unit includes the following parts:

(1) Biomass feeding system;
(2) Air supply, control, and preheating;
(3) FBG reactor;
(4) Gas sampling and offline analysis section;
(5) A cyclone filter for ash removal;
(6) A heated high-performance ceramic wall filter, where about 99% of the fine gas particle charge is maintained;
(7) A water purification unit, consisting of three refrigerants and a condenser in which the gas tar content is minimized.

The max capacity and efficiencies of the SMARt-CHP bioenergy generation technology are:

- Max. capacity (t/y) = 187.5
- Electrical efficiency, n_e % = 25
- Thermal efficiency, n_{th} = 67

3.3. Choice of Experimental Data

The experimental data on which the simulation was based were derived from experiments performed at our laboratory in the temperature range of 700–900 °C, with under-stoichiometric conditions of oxygen supply. A sub-stoichiometric ratio of 10/90

v/v % O_2/N_2 was used. The conversion yield reached 24% wt./wt. The heat produced was on average 10.6 MJ/Nm^3. The syngas composition mainly consisted of carbon monoxide (CO), carbon dioxide (CO_2), methane (CH_4), hydrogen (H_2) and traces of ethylene (C_2H_4) and ethane (C_2H_6) with average values of 30, 40, 10, 16, 0.75, and 1.15% in v/v, respectively. Char, the solid gasification by-product, yielded at 33.5 wt.% [6].

3.4. Choice of Simulation Modeling

Although models can mimic many natural phenomena, they require very detailed information (geometry, materials, and boundary conditions) and high computational resources. Models are classified as stoichiometric and non-stoichiometric depending on whether they are based on equilibrium constants or minimizing Gibbs free energy. Non-stoichiometric equilibrium models are the most common approach to describing the performance of a Fluidized Bed Gasifier (FBG). Aspen Plus is used to model biomass gasification processes [13].

ASPEN Plus is the chemical industry's leading process simulation software that allows the user to build a process model and then simulate it using complex calculations (models, equations, math calculations, regressions, etc.), while it enables lifecycle modeling from design through operations combining accuracy and time-saving. It is being used by many researchers to simulate the gasification process of biomass and wastes [4,14].

Ultimately, the choice of the model largely depends on the targets and experimentally available information. We knew that Aspen Plus modeling involving an FBG could be difficult due to the complexity of the hydrodynamic liquefaction and the complex nature of the natural and chemical phenomena that occur within the FBG.

Hypotheses and Model Assumptions

The accuracy of simulation results strongly depends on the decisions and assumptions that have been made. Table 2 presents the assumptions made in this study for simulating the process of FBG.

Table 2. Model's hypotheses and assumptions.

1	The process is stable and isothermal (heat losses are zero).
2	The gasifier is in steady state with uniform temperature and pressure.
3	The method used for simulation is "ideal".
4	All gases are considered ideal (O_2, N_2, H_2, CO, CO_2, CH_4 and H_2O).
5	All gases are evenly distributed in the reactor.
6	The reactions are in chemical equilibrium and react quickly.
7	Hydrodynamic equations in the Fluidized Bed Gasifier (FBG) were not taken into account.
8	Sulfur and nitrogen reactions have not been considered.
9	Char contains only carbon.
10	Ash in biomass is considered an inert material.
11	Biomass particles have temperature uniformity (temperature gradient = zero).
12	Pressure in the gasification furnace is constant and equal to the atmospheric.
12	Drying phase is avoided due to low moisture content of the feedstock.
14	Tar defined as "C_6H_6" (same thermochemical properties of benzene).
14	Char defined as carbon with the thermochemical properties of graphite.

3.5. Choice of Processes

The processes of this simulation concern pyrolysis, combustion, and gasification as well as the cleaning of the gaseous product. The first stage involves pyrolysis, which simulates the thermal decomposition of biomass before oxidation (i.e., the gasification zone of the gasifier, where the biomass is broken). The pyrolysis process is achieved at high temperatures around 500 °C and its goal is the conversion of biomass from non-conventional to simple components (H_2, CO, CO_2, CH_4 and H_2O). The second stage concerns combustion and gasification (i.e., the combustion zone and gasification zone of the gasifier, where the conventional components react with the gasification agent to further oxidation–reduction reactions after preliminary gasification).

For the evolution of temperature, we used information provided by the experimental study. First, the fuel was loaded into the reactor from the top at room temperature, while the gasification agent was introduced from the bottom of the reactor at ambient temperature if it was air and at above 100 °C if it was steam. As biomass moves downwards, it is subjected to cracking, carried out at a temperature up to 500 °C. Then, the gasification stage takes place, in a temperature range of 550–900 °C. The combustion products introduced into the reactor in the oxidation zone can rise the temperature up to 1100 °C for the need of breaking down the heavier hydrocarbons and tar of the syngas. As these products move downwards, they enter the reduction zone where a production gas is formed by the action of carbon dioxide and water vapor. Hot and dirty gas passes through a system of refrigerators, cleaners, and filters before being sent to engines, as it is the standard way [15].

In the present simulation in the first-round calculations, air was used as the gasification agent, the oxygen of which, in combination with the high temperature, leads to combustion. At the same time, the remaining conventional components and combustion products were led to the gasification stage where the achieved temperature was above 700 °C. In this process, reactions such as the methane reforming reaction (MSR) and the water–gas displacement reaction (WGS) play an important role in the production of the high-value gas product based on the Gibbs free energy minimization principle.

The final stage involves wet cleaning through cooling water and the separation of clean gas and unwanted liquid products.

3.6. *Choice of Reactions System*

Biomass contains carbon, hydrogen, and oxygen as the main chemical components. Therefore, it can be represented by the molecular formula CxHyOz which can be quantified by the final analysis, where x, y and z represent the elemental fractions C, H, and O, respectively. RH molecular form is described as $CH_{\alpha h}O_{\beta h}(SiO_2)_{\delta h}$, where ah, bh, dh and a, b, d were calculated by the analysis of RH from Table 1.

We also assumed that the RH char has the chemical formula $CH_{\alpha}O_{\beta}(SiO_2)_{\delta}$.

The homogeneous and heterogeneous chemical reactions that we considered to occur in the gasification process are shown in Table 3. The number next to the reactions indicates the order in which they are performed. Hydrogen and carbon in combustion reactions (R-3, R-2), as well as water–gas in displacement and methanization reactions (R-7, R-4) are all exothermic and ideally provide the system with the required energy. On the other hand, steam reforming, Boudouard and water–gas shift reactions (R-6, R-8, R-5) are endothermic and their effect on gasification products becomes more apparent at high temperature.

Table 3. Reactions used in simulation with Aspen Plus.

Reaction	Reaction Type	No
$CH_{ah}O_{\beta h}(SiO_2)_{\delta h} \rightarrow CH_aO_\beta(SiO_2)_\delta + \text{volatiles}$	Decomposition	R-1
Homogeneous Reactions		
$H_2 + 0.5O_2 \leftrightarrow H_2O$	Hydrogen combustion	R-3
$CH_4 + H_2O \leftrightarrow CO + 3H_2$	Steam reforming	R-6
$CO + H_2O \leftrightarrow CO_2 + H_2$	Water–gas shift	R-7
Heterogeneous Reactions		
$CH_aO_\beta(SiO_2)_\delta + \gamma O_2 \leftrightarrow \left[2 - 2\gamma - \beta + \frac{\alpha}{2}\right]CO + \left[2\gamma + \beta - \frac{\alpha}{2}\right]CO_2 + \left(\frac{a}{2}\right)H_2O + \text{ash}$	Combustion	R-2
$CH_aO_\beta(SiO_2)_\delta + \left[\frac{(4-a+2b)}{2}\right]H_2 \leftrightarrow CH_4 + \beta H_2O + \text{ash}$	Methane formation	R-4
$CH_aO_\beta(SiO_2)_\delta + (1\text{-}\beta)\ H_2O \leftrightarrow CO + \left[1 - \beta + \left(\frac{\alpha}{2}\right)\right]H_2 + \text{ash}$	Water–gas	R-5
$CH_aO_\beta(SiO_2)_\delta + CO_2 \leftrightarrow 2CO + \beta H_2O + \left[\left(\frac{\alpha}{2}\right) - \beta\right]H_2 + \text{ash}$	Boudouard Reaction	R-8

This simulation involves 3 stages:

- ❖ The first stage involves the pyrolysis reaction, which is represented by the R-1 reaction and the volatiles include CO, CO_2, H_2, H_2O, CH_4, hydrocarbons with low molecular weights such as C_2H_4 and C_2H_6, carbon and tar.
- ❖ The second stage is combustion, where the RH char ($CH_\alpha O_\beta(SiO_2)_\delta$) is first subjected to partial combustion with oxygen to produce CO and CO_2, represented by the R-2 reaction, according to literature [4].
- ❖ The third stage consists of gasification reactions represented by gasification reactions R-3 until R-7, according to the literature [4].

In fact, there is another preliminary stage before cracking, which is the drying phase to reduce the raw material moisture content below 10 wt.%., depending on the moisture content of the raw material. We neglected the drying stage in this study because RH' moisture content is 9.5 wt.% < 10 wt.% [6].

We also considered the process of gasification to take place at the atmospheric pressure that is the most common [16].

3.7. Choice of Reactor Blocks

The simulation of the gasification reactor was performed in Aspen Plus software with the array of 2 reactors, each of which had a separate use which at the same time led to the result. For a multi-phase or multi-action system such as RH gasification, which involves multiple decompositions, combination, and adverse reactions, it is recommended to use the type of Gibbs reactor (R-Gibbs) created in Aspen Plus required to solve all of them to predict equilibrium compositions. This type of reactor is based on minimizing the total Gibbs energy of the mixture products and allows control and transport.

Since R-Gibbs cannot handle non-conventional components such as RH, in the case that some electricity or heat is needed, this can be inserted into the R-Yield block. In this block, RH is converted into a system of equivalent environmental components at the same levels of enthalpy. This current, generated after R-Yield, in combination with the air required for partial combustion and gasification, is directed to the R-Gibbs block to produce the products of the gasification reactions. The R-Gibbs subunit calculates adiabatic reactivity temperatures, such as the equilibrium component (estimated using Gibbs free energy minimization). The R-Gibbs calculation subunit can also be used when one or more reagents are not fully involved in equilibrium conditions. This is achieved by specializing in the extent of equilibrium for the ingredients.

In the case of the gasification of RH where there are adiabatic conditions, the equilibrium of the composition of the product provided by R-Gibbs depends on the flow rates, composition, and temperature of the surface materials (rice husk and air) supplied to the gasifier. The reactor blocks are presented in Table 4.

Table 4. Reactor models used in this simulation in Aspen Plus.

Reactor	Name	Description	Input Data
RGIBBS	Balance reactor	Calculation based on the minimum free energy of GIBBS due to the limitation of individual equilibrium	Pressure, temperature
RYIELD	Performance reactor	Calculation of the chemical reactor of the ratio of distribution of known reaction products and the unknown kinetic model	Pressure, temperature

3.8. Flow Sheet of Air Gasification

In the software, the biomass supplied to the gasifier is characterized by the ultimate and proximate analysis and not by its chemical formula, as it is classified as non-conventional. The HCOALGEN and DCOALIG tool was used, for the the final analysis and sulfur analysis, to calculate the lowest heating value (LHV), the enthalpy calculation (HCOALGEN) and the density (DCOALIG) of the biomass (non-conventional component). The Peng Robinson equation was used to estimate all the physical properties of conventional components produced by the gasification process.

The R-Gibbs block calculates the equilibrium of the chemical equilibrium and the phase by minimizing the Gibbs free energy of the system. Before feeding the biomass to the R-Gibbs block, it must be decomposed into conventional elements using the R-Yields reactor. Thus, the R-Gibbs block was used to precisely simulate oxidants and reduction zones in the gas reactor. A mixer block was used to mix the products of the R-Yield reactor (Decomp) with the flow of air, in a sub-stoichiometric quantity, before entering the R-Gibbs block.

Figure 1 presents a comprehensive Aspen Plus flow sheet for the fluidized-bed gasification process, while Table 5 descripts the Aspen Plus reactor blocks considered in the model.

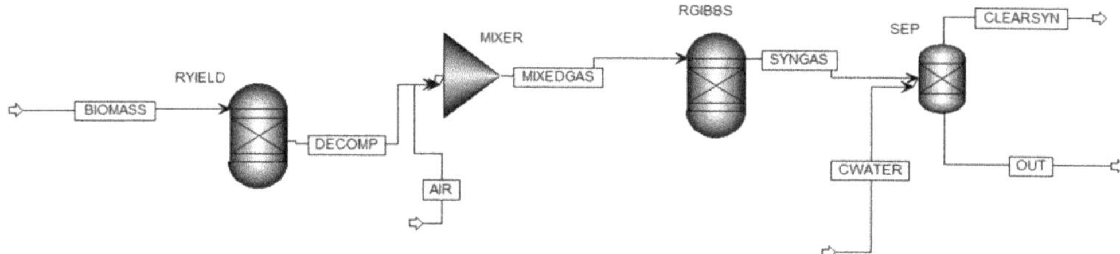

Figure 1. Comprehensive Aspen Plus flow sheet for the fluidized-bed air gasification.

Table 5. Reactor models used in this simulation in Aspen Plus.

Aspen Plus ID	Block ID	Description
RYIELD	DECOMPOSER	Converts nonconventional biomass into conventional components
RGIBBS	GASIFIER	Minimizes Gibbs free energy to reach chemical equilibrium and calculate outlet composition
MIXER	MIX	Mixes the different outlets of the blocks to reproduce a single product steam
SEP	WSEP	Separates tar and H_2O from the rest of the reaction steam

The performance of the individual products of the R-Yield block can be estimated using a "block calculator", i.e., a subroutine written in Fortran language defined by the user to estimate the performance of volatility products based on the final and immediate analysis of the biomass, or with approximate models of the reactions that take place during the firing stage and are processed in Excel software.

3.9. Energy Balances

We considered that the mixing ratio is high due to the fluidized bed gasifier, and that removal, combustion, and all gasification processes occur at a high rate, at the operating temperature. Therefore, the mass balance can be given by the Equation (1), considering the Gibbs free energy minimization concept.

$$\begin{aligned} Q_C(T,P)|_{\text{combustion}} &= -Q_C(T,P)|_{\text{heat loss}} + Q_{rh}(T,P)|_{\text{rice hush}} + Q_O(T,P)|_{\text{oxidant}} + Q_d(T,P)|_{\text{drying}} \\ &+ Q_{dv}(T,P)\Big|_{\text{devolatilization}} + Q_g(T,P)\Big|_{\text{gasification}} - Q_c(T,P)\Big|_{\text{elutriation}} - Q_p(T,P)\Big|_{\text{product gas}} \end{aligned} \quad (1)$$

where:

$$\left(dG_{system}\right)_{T,P} = 0 \quad (2)$$

$$nG = \Sigma\left(n_i \Delta G^0_{fi}\right) + (\Sigma n_i) RTlnP + RT\Sigma\left(n_i ln_{yi}\right) + RT\Sigma\left(n_i ln_{\Phi i}\right) \tag{3}$$

or

$$min \frac{G}{RT} = \sum_{i=1}^{N} n_i \frac{\Delta G^0_{f,i}}{RT} + lnn_i \sum n_i \tag{4}$$

The term "elutriation" in Equation (1) refers to the separation of fine particles from smaller ones. It is important to include the contribution of the elutriation of fine particles from fluidized beds because it affects the residence time, conversion, and is used for reaction, drying and in energy balance.

The above Equation (4) is the total Gibbs equation (*nG*) that must be minimized depending on the composition of the individual compounds at the operating temperature and pressure. This depends on the constraints imposed by individual balances written on closed systems [4].

3.10. Air Gasification Syngas Composition Estimation

We based the estimation of syngas composition on the main reactions that took place during pyrolysis. Essentially, we followed the procedure described elsewhere [17] but with a consideration of carbon efficiency close to 100% (YC = 0).

Therefore, based on the proportions resulting from the above reactions, the simplified yields of the conventional products follow the relationships:

$$Y_{CO2} = 2 * Y_{CO}, \tag{5}$$

$$Y_{CH4} = 0.3 * Y_{H2}, \tag{6}$$

$$\Upsilon_{char} = 0.35 \tag{7}$$

$$Y_{CxHy} = 0.03 \tag{8}$$

Combining the above and the using the $\sum_{1}^{n} Y = 1$ subroutine, we calculated the yields (v.v%) of CO, CO_2, CH_4, H_2, C_2H_4, C_2H_6, and H_2O.

The flows of the material were as follows: The releases of biomass gas components such as CO, H_2, CO_2, CH_4, N_2, H_2O, and O_2 are defined as routine components. Biomass is the non-conventional ingredient. The decomposition unit is very similar to the R-Yield performance reactor. In this section, the biomass is decomposed into some conventional solid elements, i.e., the gasification product in a simpler form of each element such as O_2, H_2, C, N_2, and ash. The energy flow in this process is as follows: Some of the heat generated by the combustion of carbon is the heat loss of the whole system, and some flows to the pyrolysis reactor. The rest of the heat is provided by the gasification reaction to create gas. In the cracking unit, the carbon conversion ratio is 99%, i.e., it approaches 100% in the gasifier.

In the R-Gibbs reactor, chemical equilibrium reactions have been tested to represent the gasification process, namely the methane reforming reaction and the water–gas displacement reaction:

Methane reforming reaction (R-6): $CH_4 + H_2O \leftrightarrow CO + 3\,H_2$, $\Delta H = 206$ kJ/mol (9)

Water-gas reaction (R-7): $CO + H_2O \leftrightarrow CO_2 + H_2$, $\Delta H = -40$ kJ/mol (10)

The methane reforming reaction is a chemical reaction that converts methane into carbon monoxide and/or hydrogen. WGS converts CO and H_2O to extra H_2 and carbon dioxide, as the reaction does not change linear sets and therefore the effect of the pressure on the reaction is minimal.

Assessing tar and char yields is a difficult task through a thermodynamic equilibrium model because tar is usually a non-equilibrium product. Since the predictions of mathematical models are substantially improved when tar formation is included, in this study, tar and carbon yields were considered as input parameters and were determined independently of the gasifier operating conditions, according to other bibliographic models. Thus, they were placed as inert ingredients in the R-Gibbs reactor. At high temperatures, such as those examined, the tar content is very moderate, while the gas efficiency is very high.

Tar was described as "C_6H_6" with the same thermochemical properties of benzene, while char was defined as carbon with the thermochemical properties of graphite.

4. Indicators for Monitoring and Assessing of FBG System

In order to be able to derive reliable and comparable results of the three scenarios, certain indicators must first be defined in addition to the gasification temperature, characterized as active evaluation indicators. Parameters such as temperature (T), equivalence ratio (ER), and biomass vapor ratio (SBR) are suitable for the synthesis of syngas, as well as the lowest heating value (LHV). The indicators used for the sensitivity analysis in this study are:

(1) Air–biomass equivalent ratio (ER).
(2) Steam–biomass ratio (SBR).
(3) Lower heating value (LHV).
(4) Higher heating value (HHV).
(5) Cold gas efficiency (CGE).
(6) Carbon conversion efficiency (CCE).

4.1. *Air–Biomass Equivalence ratio (ER)*

In this simulation, the gasification process was investigated by changing the air flow and consequently oxygen, which affects the ER equivalence ratio that is the main operating parameter. ER is defined as the air-to-biomass weight, relative to the stoichiometric air-to-biomass weight required for complete combustion. The ratio of air used in the system to the stoichiometric required air (ratio of air equivalence ratio) is an important factor to consider. Its wise choice discourages the stimulation of oxidation reactions. The reason for combustion equivalence is the ratio between the available oxidizer and the stoichiometric quantity required for the complete reaction. It will have a value of 1 for full combustion and 0 for pyrolysis, while the appropriate values fall in the range 0.19–0.43. Oxygen availability, both as a free molecule and as a percentage in the water molecule, is a key factor in gasification. ER is defined by the following equations [14]:

$$ER(O_2) = \frac{\frac{\text{air used(kg)}}{\text{biomass used(kg)}}}{\frac{\text{stoichiometrically demanded air(kg)}}{\text{biomass used(kg)}}} = \frac{\text{actual air to biomass ratio}}{\text{stoichiometric air to biomass ratio}} = \frac{\text{feed } O_2\left[\frac{kg}{s}\right]}{\text{flow of } O_2 \text{ for complete combustion}\left[\frac{kg}{s}\right]} \quad (11)$$

4.2. *Steam to Biomass Ratio (SBR)*

In a steam gasification scenario, steam is used as the oxidizing agent instead of air. In this case, the steam-to-biomass ratio (SBR) is used as the ratio between the flow rate of the incoming steam to the flow rate of the biomass fed, as can be seen the following equation [18]:

$$SBR = \frac{\text{steam mass flow (kg)}}{\text{biomass feed rate(kg)}} \quad (12)$$

The biomass feed rate is maintained as constant while the steam flow is varied. Therefore, it is clear to foresee that above the SBR optimum range, the gas yield, LHV, and carbon conversion efficiency will tend to decrease because high amounts of unreacted H_2O will appear in the syngas, causing thermal efficiency to decline significantly. The optimum range of SBR is 0.2–0.4, based on the bibliography [14].

4.3. *Lower Heating Value (LHV)*

The lower heating value (LHV) is defined as the net calorific value and is determined by subtracting the heat of vaporization of water vapor. The main research goal is to produce gas enriched in CO, H_2, and CH_4 because the presence of these fuels leads to gas of high heating value, suitable for further exploitation in internal combustion engines and turbines for power generation. The lowest heating value (LHV) of the produced gas is calculated using the following equation [14]:

$$LHV_{syn} = \frac{(30 * X_{CO} + 25.7 * X_{H_2} + 85.4 * X_{CH_4}) * 4.2}{1000}, \ MJ/Nm^3 \quad (13)$$

where X_{CO}, X_{H2}, and X_{CH4} are the linear fractions of the gaseous products in syngas.

4.4. Cold Gas Efficiency (CGE)

CGE is the key index that measures the global performance of the gasification process. It is defined based on the first law of thermodynamics as the ratio between the chemical energy of raw syngas (calculated as the product of syngas mass flow and its lower heating value) and the chemical energy of RH feedstock. Therefore, CGE is the energy output over the potential energy input (chemical energy contained in the product gas with respect to the energy contained in the initial solid fuel) based on the LHV of both the solid fuel and the product gas. The CGE indicates the percentage of energy content of RH inherited from the syngas and can be calculated from the following equation [18]:

$$CGE = \frac{LHV_{gas} * V_{gas}}{LHV_b * F_b} \quad (14)$$

where LHV_{gas} is the producer gas's lower heating value, V_{gas} is the volume of produced gas and LHV_b is the lowest heating value of rice husks which is equal to 7.13 MJ/Nm^3. F_b is the RH feed.

4.5. Carbon Conversion Efficiency (CCE)

Carbon conversion efficiency (CCE) expresses how much of the natural carbon from biomass waste is transported to the gas produced. The equation used to calculate CCE is shown below [18]:

$$CCE = \frac{\text{total carbon outlet syngas} * 100}{\text{carbon in biomass feed}} = \frac{[12 * V_{gas}(CO\% + CO_2\% + CH_4\% + 2 * C_2H_X\%) * 100\%}{C_b\% * 22,4} \quad (15)$$

where V_{gas} is the volume of produced gas, CO $v/v\%$ the volume percentage of gaseous species in the producer gas and C_b is the wt.% of carbon in the RH feed.

5. Results: Model Validation with Experimental Results (ERes)

The simulation results (SRes) for the FBG gasifier model were validated through comparisons with experimental data from one previous study [6]. In order to be able to compare data obtained from the simulation of Aspen Plus, the gasification conditions of the experiments must first be provided. In each gasification cycle in the experimental study, 5 g of biomass of rice husk biomass were fed to the gasifier, so a 0.005 t/cycle. Additionally, as a gasification agent, the air under stoichiometry of 10/90 $v/v\%$ (O_2/N_2) with a flow of 200 mL/min was used and the residence time in the gasifier was 32 min on average [14]. Therefore, with simple calculations for each gasification cycle, 0.11 L/cycle of the gasification agent was estimated.

Table 6 shows the simulated results (SRes) of the syngas composition for three air gasification temperatures (T = 700, 800, 900 °C) and for various experimental results (Eres).

Table 6. Simulated results (SRes) of the syngas composition with temperature and experimental results (ERes).

ER	CO *v*/*v*%	H_2 *v*/*v*%	CH_4 *v*/*v*%	CO_2 *v*/*v*%	CCE	LHV (MJ/Nm^3)
			T = 700 °C			
0.25	55.9	18.8	20.2	4.95	0.94	16.4
0.35	60.6	18.8	13.4	7.13	0.86	14.2
0.55	62.7	18.1	4.74	14.4	0.72	11.4
0.75	57.2	16.4	0.96	25.4	0.57	9.3
1.30	47.2	14.3	0.36	38.1	0.48	7.5
1.60	28.6	9.74	0.02	61.5	0.20	4.7
			T = 800 °C			
0.25	65.8	19.8	14.2	0.73	1	15.6
0.35	73.6	19.3	5.94	1.25	0.98	13.4
0.55	72.1	18.9	0.49	8.43	0.83	11.3
0.75	63.1	16.7	0.09	20.1	0.63	9.7
1.30	54.3	14.5	0.03	31.2	0.53	8.4
1.60	34.5	9.72	0.01	55.8	0.28	5.4
			T = 900 °C			
0.25	67.1	20.1	12.9	0.04	1	15.4
0.35	73.8	20.4	4.32	0.25	1	13.2
0.55	74.5	19.1	0.05	6.34	0.90	11.4
0.75	66.9	16.8	0.01	16.2	0.69	10.2
1.30	59.3	14.6	0.002	26.1	0.60	9.1
1.60	41.4	9.71	0.001	48.9	0.38	6.1

The ER value is directly related to the oxygen/air content in the gasifier, and if it is high, it can turn the gasification process towards combustion as Table 6 shows. Higher ER values lead to a decrease in syngas heating value and in the higher conversion of H–Cs to CO and CO_2, a decrease in tar yield and CH_4 content in the syngas. Increasing temperature increases H_2 production in product gas due to the gasification of char and methane reforming reactions (Figure 2). Figure 2a–d compare the SRes with ERes of syngas composition in CO, CO_2, CH_4 and H_2 in function of the air gasification temperature.

As it can be noticed in Figure 2, SRes show good accuracy in the prediction of H_2 but not for CO, CO_2 and CH_4, content. This indicates that the model needs modification to improve the accuracy of prediction.

This can be attributed to the fact that the RGIBBS reaction simulates better an entrained flow gasifier and not so well a fluidized bed gasifier, for which a semi-empirical model might fit better than the RGIBBS reaction. For a more detailed calculation of the difference between SRes and ERes, Table 7 presents the calculated deviations by using the following equation: % deviation = [(SRes) − (ERes)/(ERes)] × 100.

Table 7. Deviation between the simulated (SRes) with experimental results (ERes) at 900 °C with ER = 0.3 (air gasification)

Syngas Composition	Simulation Results (SRes)	Experimental Results (ERes)	% Deviation [(SRes) − (ERes)/(ERes)] × 100
H_2	13	16	−18.7
CO	43	35	+22.3
CO_2	46	38	+21.0
CH_4	3	9.7	−60.0
LHV	7	10.5	−33.3
CGE	37	21	+76.2
CCE	17	26	−34.6

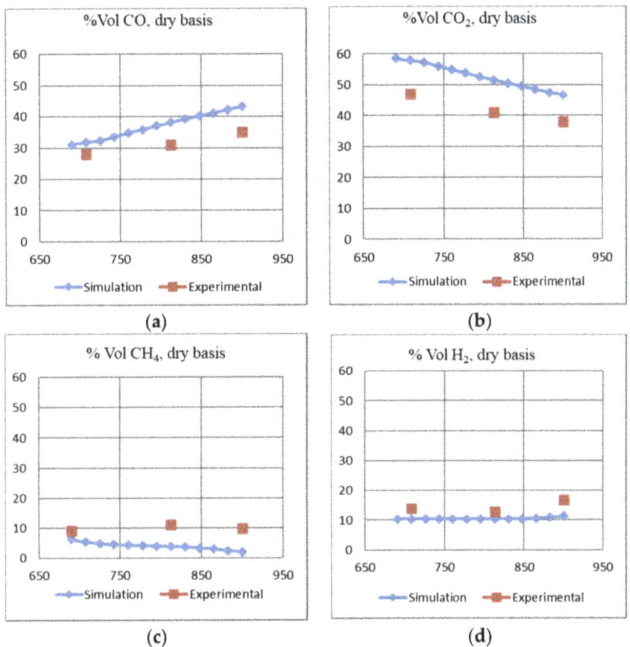

Figure 2. Comparison of simulated and experimental data: effect of air gasification temperature on syngas composition: (**a**) CO; (**b**) CO_2; (**c**) CH_4; and (**d**) H_2.

As it can be noticed in Table 7, deviations vary from −60.0 to +76.2. These are the lower and upper deviations that mainly occur in the case of CH_4 and consequently to the cold gas efficiency (CGE). While experimentally it appears that in the syngas there is a certain amount of methane, this is not the case in the simulated results. This devaluation of CH_4 is due to the minimization of Gibbs energy and the ideal chemical equilibrium reactor that were hypothesized in the simulation, which do not occur in real commercial gasification systems.

Additionally, the SMARt-CHP technology that was considered in this study for the experimental results produces tar and hydrocarbons (mainly methane), components that were neglected in the equilibrium-based predicted model.

Similarly, some small differences in the composition of the gaseous products are due to the consideration of the R-Yield reactor to simulate gasification in Aspen Plus. The RGIBBS reaction is rather closed to the entrained flow gasifier and not to the fluidized bed gasifier, for which a semi-empirical model might fit better than the RGIBBS reaction.

6. Sensitivity Analysis for Monitoring and Assessment by Using Indicators

A sensitivity analysis was conducted to monitor and assess the studied system by using the indicators described in the previous chapter.

6.1. *Effect of Equivalence Ratio (ER) and Gasification Temperature on Syngas Composition*

During gasification, emphasis is placed on the maximizing gas efficiency to produce a gas with an HHV to be efficient and used to generate electricity. Two parameters are the main ones that affect the efficiency and composition of the gas:

(1) ER.
(2) Temperature.

If a high ER is used, the syngas content on CO, H_2 and CH_4 decreases with a higher ER (Figure 2); and the gas LHV decreases. At the same time, increasing the ER allows to increase the temperature of the reactor, promoting a higher flow through the reactor, and reducing the tar content in the syngas. According to the above figure, the concentration of CH_4 in syngas decreases dramatically with increasing ER. The CO and H_2 content decreases with very high ER. For the above reasons, an accurate choice of the two parameters is necessary to optimize the process.

At the Aspen Plus simulation of air gasification, the ER ratio was set at 0.3 to achieve a syngas with a high heating value. Figure 3 depicts the effect of ER on the composition of the syngas simulated results (SRes).

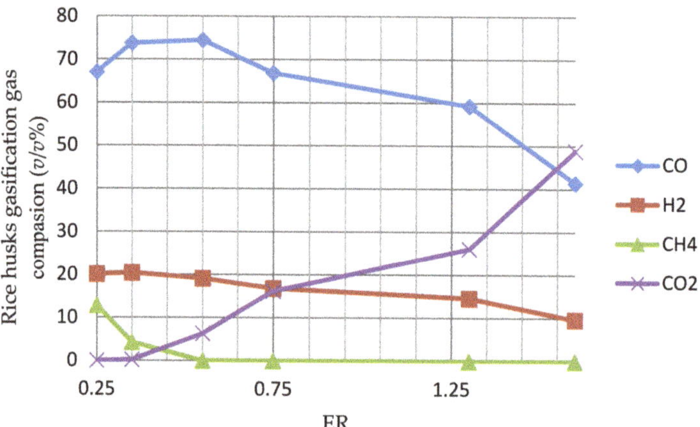

Figure 3. Effect of ER on the composition of the syngas simulated by Aspen Plus.

Based on the results from the above ER comparison chart, it is observed that for ER > 0.3, carbon dioxide (CO_2) increases sharply, which is not desirable. Therefore, in the case of rice husk air gasification, an ideal ER ratio is proved to be 0.3 (quite close to the bibliographic one which is 0.27). It is obvious that for a given temperature, the increase in air (higher ER) leads to a decrease in the final efficiency of syngas.

Further increase in ER leads to reductions in CO and H_2 concentrations, which is probably due to the favorable combustion reaction. The CO_2 concentration increases sharply with the increase in ER due to complete combustion and reaches a value in the range of 20–30% at ER = 1. The change in the concentration of CH_4 with an increase in ER is considered negligible. Further increases in ER were found to lead to reductions in CO and H_2 volume fractions due to combustion reactions. LHV increases with increasing ER to the value in the range of 0.35 and then begins to decrease dramatically. In conclusion, ER had the opposite effect on LHV from temperature, i.e., higher ER reduced LHV_{gas} due to the oxidation of part of the gaseous gases present in the syngas.

Moisture content (MC) of biomass affects the efficiency of the gasification process. It is known that the high content of MC is responsible for reducing H_2 and CO in gas production and increasing CO_2. As a result, the heating value of syngas decreases while the MC increases. For this reason, in this simulation, the biomass of rice husks with moisture content below 10% was used as a raw material.

The gasification unit was simulated in Aspen Plus software in the temperature range of 700–900 °C, with an air gasification agent and with stoichiometry (10/90 v/v% O_2/N_2). This temperature range was implied by the experimental data because the process was studied in the temperature range of 550–900 °C to optimize the syngas quality. The ER was set to 0.3, a value set in the experimental study. The effect of temperature on the quality and the energy efficiency of the syngas was studied in Aspen Plus software and is shown in Figure 4.

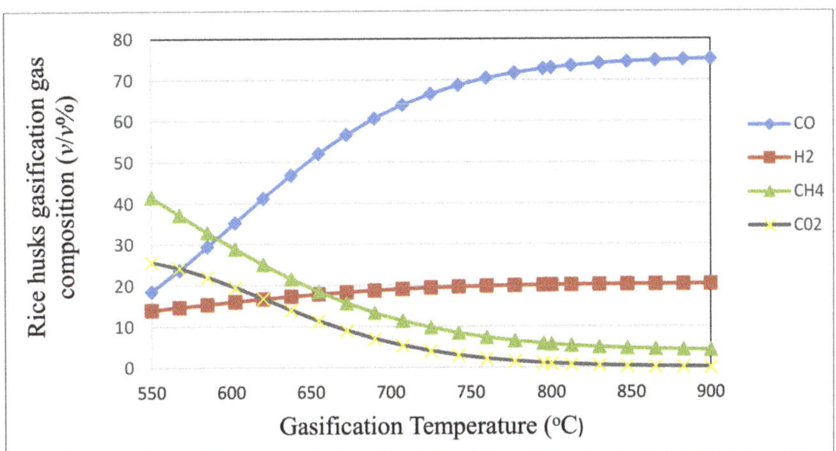

Figure 4. Effect of temperature on the syngas composition simulated results of air gasification of rice husks in Aspen Plus (free of H_2O and N_2).

The temperature of the gasifier affects the overall composition of the final product as shown in the diagram above. This is because some of the related chemical reactions that take place inside the gasifier are endothermic. Higher temperatures favor endothermic reaction products according to Le Châtelier's principle. Temperature promotes the formation of a gas produced with higher H_2 and CO contents and therefore higher LHV. On the other hand, the content of CH_4 and CO_2 follows an opposite trend. CH_4 decreases with temperature because the methane reaction formation is exothermic.

According to the above Figure 4, CO_2 follows a downward trend until it is eliminated as the gasification temperature increases, in contrast to CO, which while initially having a lower composition than CO_2, follows an upward trend reaching very high percentages. H_2 shows a relatively small increase and stabilizes at 20 v/v% from 750 °C onwards. CH_4 shows a very downward trend and stabilizes at 6 v/v% from 750 °C onwards. Finally, it should be noted that the remaining hydrocarbons (C_2H_4 and C_2H_6) in the whole range of temperatures have a composition below 1 v/v%.

The reduction in CO_2 concentration could be attributed to the Boudouard reaction which takes place at a higher temperature range compared to the water–gas shift reaction. Therefore, CO production and CO_2 consumption are preferred. In addition, methane reforming reactions affect the CH_4 concentration which is reduced to a higher gasification temperature. The bottom line is that the produced gas from the simulation of the Aspen Plus gasification unit is rich in CO and H_2, but poor in CH_4 and CO_2.

The molecular weight of the produced gas is 22.

6.2. *Effect of Gasification Temperature on Syngas Low Heating Value (LHV)*

It was observed that this LHV of the syngas stabilized at 13 MJ/Nm^3 from 850 °C onwards. It is considered that at 850 °C, the gasifier reaches the highest fuel conversion. During these calculations, the LHV values of the syngas at 700, 800, and 900 °C were taken to be around 14.5, 13.0 and 13.0%, respectively.

6.3. *Effect of Gasification Temperature on Cold Gas Efficiency (CGE)*

CGE indicates the percentage of energy content of RH transferred in the gas product. CGE for all raw materials is directly proportional to the gasifier temperature according to the definition and Equation (15). However, gas LHV decreases with temperature, and CGE is higher at a higher gasification temperature due to the increased volume of gas product. The CGE from the rice husks is maximized at 850 °C where the gasifier reaches the highest fuel conversion. During these calculations, the CGE values of the FBG gasifier at 700, 800, and 900 °C were taken to be 85.0, 92.0, and 90.0%, respectively. The high CGE suggests that the coke is cracked.

6.4. Effect of Gasification Temperature on Carbon Conversion Efficiency (CGE)

Carbon conversion efficiency (CCE) expresses how much of the natural carbon from rice husk waste is transported to the produced gas. The maximum efficiency of carbon conversion (CCE) is at 900 °C, where the gasifier reaches the highest fuel conversion, and it is equal to 22%, while at 850 °C it reaches the 21% conversion. During these calculations, the CCE values of the FBG gasifier at 700, 800, and 900 °C were taken to be 16.5, 20.0, and 22.0%, respectively.

6.5. Study of Alternative Gasification Scenarios

The oxidizing agent has a significant effect on the heating value syngas produced. However, the main scenario studied was that of air gasification, and simulations of other two scenarios were attempted by using all the same hypotheses and conditions whilst only changing the gasification agent. Thus, the second scenario simulated was the steam gasification in the R-Gibbs reactor and the third scenario was the (air + oxygen) gasification.

The flow sheet of the steam gasification scenario is presented in Figure 5. The flow sheet of the (steam + oxygen) gasification is depicted in Figure 6.

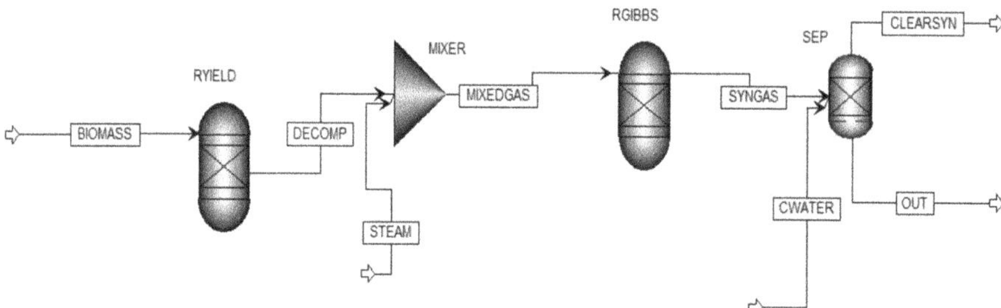

Figure 5. Aspen Plus flow sheet for steam FBG scenario.

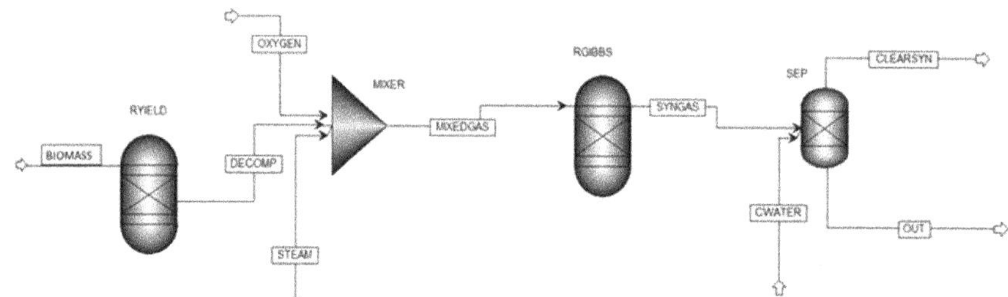

Figure 6. Aspen Plus flow sheet of (oxygen + steam) FBG scenario.

The results of the second alternative scenario calculated by the Aspen Plus worksheet are shown in Table 8 with respect to SBR or ER, LHV and CCE indicators.

Table 8. Effect of steam-to-biomass ratio (SBR) and gasification temperature on the syngas composition simulated results (SRes) by Aspen Plus.

SBR	CO *v*/*v*%	H_2 *v*/*v*%	CH_4 *v*/*v*%	CO_2 *v*/*v*%	CCE	LHV (MJ/Nm^3)
			T = 700 °C			
0.2	52.8	21.5	25.2	0.4	1	18.1
0.35	62.3	22.7	13.9	1.0	1	15.3
0.5	67.6	23.4	6.5	2.5	1	13.2
0.75	66.8	23.6	2.1	7.5	1	11.6
1	62.6	23.4	0.9	12.9	0.95	10.7
2	47.7	22.6	0.2	29.5	0.78	8.5
			T = 800 °C			
0.2	43.4	20.6	32.3	3.6	1	19.3
0.35	49.5	21.5	23.1	5.9	1	16.8
0.5	52.0	21.9	17.1	8.9	0.96	14.6
0.75	52.6	22.3	10.7	14.4	0.95	12.8
1	50.1	22.3	7.5	20.1	0.85	11.4
2	39.8	22.1	1.8	36.3	0.62	8.10
			T =900 °C			
0.2	53.9	21.6	24.3	0.034	1	17.8
0.35	64.9	22.9	12.0	0.108	1	14.9
0.5	72.5	23.8	3.1	0.561	1	12.8
0.75	70.6	23.9	0.29	5.172	1	11.5
1	66.2	23.6	0.12	10.05	1	10.9
2	52.5	22.9	0.02	24.57	0.91	9.1

As can be noticed in the case of steam gasification, there is a larger initial amount of methane (CH_4) and less carbon monoxide (CO) compared with the results of air gasification as presented in Table 6. Finally, based on Table 8, there is an improvement in the composition of the gaseous product to the SBR = 0.5 and then as the SBR increases, the carbon dioxide increases, which is not desirable. Therefore, in the case of this study, the ideal value for the SBR was calculated as 0.4 (quite close to that of the literature which is 0.35).

The use of steam as a gasifier increases the partial pressure of H_2O in the gas reactor that favors water–gas, water–gas displacement and vapor reactions, leading to an increase in H_2 and CO_2 and a decrease in CO production as SBR increases. The heating value and hydrogen content of syngas are generally higher when the gasification of RH occurs with steam than when it occurs with air. However, based on Table 8, the results are almost similar to those presented in gasification with air factor (Table 6).

Both gasification agents (air and steam) are efficient with the only difference in the case of air, however, it is a cheap agent as opposed to the steam and steam gasification needs more energy to turn water into steam to be used in the process, although in the case of steam gasification, syngas has a higher H_2 composition, resulting in higher LHV value.

Regarding the third alternative scenario of gasification with (steam + oxygen), it can be noticed that the best results are derived when ER = 0.3 and SBR = 0.4.

The comparison of the scenarios based on the syngas composition is depicted in Table 9.

Table 9. Composition for syngas of (steam + oxygen) gasification for ER = 0.3 and SBR = 0.5 at 900 °C.

Temperature (°C)	CO *v*/*v*%	H_2 *v*/*v*%	CH_4 *v*/*v*%	CO_2 *v*/*v*%	CCE	LHV (MJ/Nm^3)
700	51.60	19.5	9.6	19.2	0.78	11.7
800	67.8	20.9	1.4	10.9	0.9	11.1
900	70.5	21.3	0.2	8.1	1	11.3

Based on Table 9, for the gasification of RH with (oxygen + steam), the results regarding the composition of syngas were slightly better than those of gasification with air (Table 6), but worse than those of steam gasification (Table 8). This can be explained by the fact that the use of the agent (oxygen + steam) reduces the residence time of the air inside the reactor, preventing the continuous reactions in the gasifier from achieving the chemical equilibrium of a substance undertaken by the model.

However, (steam + oxygen) gasification needs an external energy source to maintain the reaction temperature, while oxygen and air are used in direct gasification because the oxidation reactions provide the energy required to sustain the temperature of the reaction. Nonetheless, oxygen is the best gasifying agent, though using oxygen is more costly and there is a risk that the gasification process may shift to combustion.

Therefore, the feasibility study that was conducted and is presented in the next chapter is the scenario of the assessment of its economic viability.

7. Feasibility Study

The Aspen Process Economic Analyzer was used for economic assessment. The SMARt-CHP characteristic values were used for the economic assessment. The cost of transportation and the price of RH was considered to be zero because it is hypothesized that the CHP unit will serve as a waste management solution for the rice processing company.

The Greek rice type "Agrino" is produced by the homonymous company which is the largest rice producer in Greece (5 t/hr). This production accounts for an RH production of approximately 20 wt.% of the total paddy weight (whole grain).

7.1. Fixed Investment Calculation (I_F)

The first step in calculating the fixed investment is to calculate the cost of mechanical equipment. Based on the calculation by the Aspen Process Economic Analyzer, the cost of equipment amounts to USD 2,279,490 = EUR 2,101,735 for the scenario of air gasification and for a capacity of 25,000 t/y rice husks (USD/EUR = 1.084).

Based on the cost of purchased equipment, the amount of fixed investment of the facility was obtained. Using the estimation method based on the cost of procurement of mechanical equipment, the amount of the fixed investment was calculated. All individual costs are expressed as a percentage of the value of the mechanical equipment and represent average values for standard chemical installations [19]. Table 10 shows fixed investment analysis using the Aspen Process Economic Analyzer. In Table 10, the percentage of the land purchasing was intentionally omitted (6%) because the gasification unit was installed in the yard of the rice processing agro-industry.

Table 10. Fixed investment (I_F) analysis based on Aspen Process Economic Analyzer.

Cost Type	Machinery (%I_F)	Cost (EUR, 2020)
	I. Direct Costs	
Machinery value	100	2,101,735
Installation	47	991,973
Control system	18	379,904
Pipelines	66	1,392,983
Electronics	11	232,164
Buildings	18	379,904
Land improvement	10	211,058
Services	70	1,477,406
Total direct investment costs	340	7,175,971

Table 10. *Cont.*

Cost Type	Machinery (%I_F)	Cost (EUR, 2020)
	II. Indirect costs	
Supervision	33	696,491
Construction	41	865,338
Total direct and indirect costs	414	8,737,800
Constructor payment	21	443,222
Contingencies	42	886,444
Fixed capital I + II	477	10,067,466
Working capital (I_W)	86	1,815,099
Total investment cost I	563	11,882,564
Cost/kw	8,000,000 kw/y	1.48

7.2. Operating Costs Estimation

Summing up all the individual expenses together with some additional ones, the total annual operating costs of the unit were obtained, as shown in the following Table 11.

Table 11. Total annual operating cost analysis (C).

Cost Type	Cost Estimation	Operating Cost (EUR, 2020)
	I. Production cost	
	A. Direct cost	
i. Raw materials		0
ii. Labor costs		564,993
iii. Supervision	15% A(ii)	84,749
iv. Utilities *		753,302
v. Maintenance/repairs	5% I_F	503,373
vi. Materials	0.75% I_F	75,506
vii. Lab expenses	10 %A(ii)	56,499
	B. Permanent cost	
i. Insurance	1 %I_F	100,675
ii. Taxes	1 %I_F	100,675
iii. Depreciation	1 %I_F	100,675
C. Additional cost	60% * [A(ii) + A(iii) + A(v)]	691,869
D. Environmental charges		32,794
Total product cost		3,065,110
	II. General expenses	
A. Administration expenses	5 % A(ii)	28,250
B. Distribution/sales costs	4% (I + II)	40,270
C. Innovation expenses	2% S	170,653
	III. Total operating cost (I+II) = 3,304,282	
i. Contingencies	2.5% III	82,607
IV. Total, C		3,386,889
EUR/t of RH		135.5

* Utilities: (a) 106,488 l/y air for the gasification; (b) 12,530,304 Whel/ y electricity for the operation of the gasification; and (c) 25,229 t/y water for the gasification products cooling.

The labor cost was calculated by using the Wessel Equation (19):

$$\frac{\text{Manhours}}{\text{days x stages}} = \alpha * \left(\frac{\text{t product}}{\text{d}}\right)^{0.24} \quad (16)$$

where α is a coefficient depending on the type of unit.

The following hypotheses were made:

- The unit is considered to be automated, and thus $\alpha = 11$;
- It was considered that the unit operates for 292 days/y, a total of 7000 hr/y; therefore, the operation coefficient λ is equal to 0.8;
- The stages of the process were considered to be three;
- The labor cost was estimated based on the price of labor–hour = EUR 14.5 [20].

7.3. Annual Sales Profits

The unit makes a profit on the one hand from the sale of electricity and heat, whilst on the other hand from the char. According to technology chosen, for an FBG unit for CHP and capacities of 100 kg/h, the energy produced is equal to 1.1–1.2 kWh for every 1 kg/h of power, regardless of the type of biomass. Thus, in our case of RH, the energy produced is set at 1.1 kWh for every 1 kg/h of RH gasified.

In the simulation performed, the capacity was 25,000 t/y RH, so by simple calculations, the generated energy was equal to around 99,000,000 kWh. From the produced energy, 28% was electricity and 72% thermal energy. Therefore, finally, 27,720,000 kWh_{el} and 71,280,000 kW_{th} will be produced by the gasification simulation unit and be sold as commodities.

The conversion of RH to char is equal to 35 wt.%. In the positive scenario of 25,000 t/y capacity, 8750 t/y char will be produced (0.35 * 25,000 = 8,750 t/y) which can be sold or used as biochar. Table 12 shows cash inflows (S) generated by the unit from the sales of the commodities.

Table 12. Cash inflows (S).

Product/Commodity	Quantity	Selling Price	Cash Inflow (EUR/y)
Rice husk char	8750 t/y	121,09 EUR/t	1,059,600
Electricity	27,720,000 kW_{hel}/y	0.101 EUR/kW_{hel}	2,799,972
Thermal energy	71,280,000 kW_{th}/y	0.065EUR /kW_{th}	4,633,200
Total annual S			**8,492,772**
Gross Income GI			**5,105,883**

The gross income of the unit is calculated by using the equation:

$$R = S - C \quad (17)$$

The assumptions made to calculate the total net revenues (NRs), are the following:

- The unit's lifespan is N = 10 years.
- Linear depreciation considered.
- Flat tax rate is t = 0.4.
- Depreciation coefficient for tax purposes is d = 1/N = 0.1.
- Depreciation coefficient for fixed capital is e = d.

The total net revenue was calculated by using Equation (18):

$$P = (R - d * I_F) * (1 - t) = 2{,}483{,}398\ €/y \quad (18)$$

7.4. Estimation of ROI and POT Indexes

The ROI index expresses the performance in relation to the amount initially invested and is calculated by using the equation:

$$\mathrm{ROI} = \frac{\mathrm{P}}{(\mathrm{I_F} + \mathrm{I_W})} = 0.21 \tag{19}$$

The POT economic index expresses the time required to equate finance with fixed investment capital and is calculated by using the equation:

$$\mathrm{POT} = \frac{\mathrm{I_F}}{(\mathrm{P} + \mathrm{e} * \mathrm{I_F})} = 2.88 \tag{20}$$

The economic indicators are very positive.

7.5. Range of Viable Capacity Estimation

Based on the economic data and by using the ROI and POT indices, we can calculate the range in which the capacity of RH that is economically viable based on sensitivity analysis by Aspen Plus software and using as parameters the capacity, the fixed investment, occupational costs, utilities, and gross profit.

The only assumption we need to keep in mind is that the ROI must exceed 0.2 and the POT must never be lower than or exceed 3.63. For this reason, the Aspen Plus software performed a sensitivity analysis on the unit's bandwidth (if the gasification unit operates for 7000 h/y). Figure 7 depicts the evolution of economic indicators with the capacity.

In conclusion, the gasification system is viable at any capacity between 25,000 and 75,000 t/y. Comparing the economic simulation results of the three gasification scenarios based on different gasification agents, we found that although oxygen-steam gasification is the most favorable option for rich syngas production, the operating costs due to oxygen and high steam requirements, render the oxygen-steam gasification the less attractive economically scenario compared to the air-gasification.

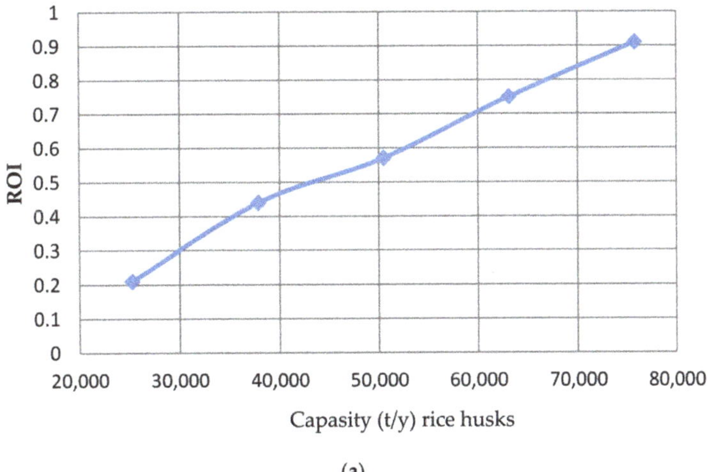

(a)

Figure 7. *Cont.*

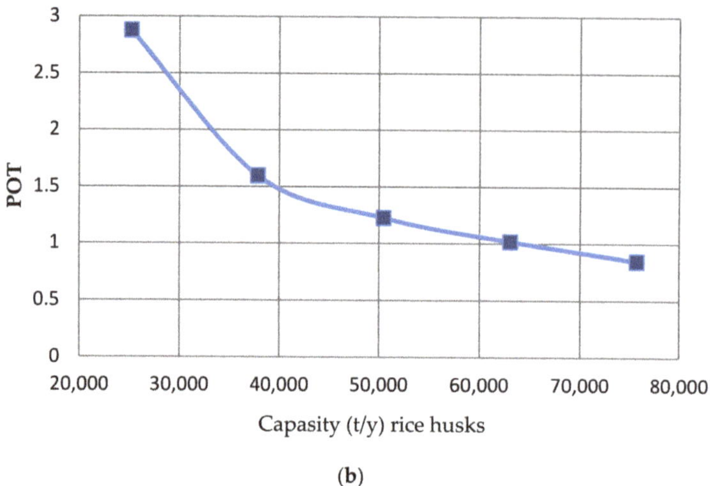

(b)

Figure 7. Evolution of the economic indicators (**a**) return of investment (ROI) and (**b**) pay out time (POT) with capacity.

8. Discussion on Environmental Issues

There are some concerns associated with RH gasification concerning the solid and gaseous by-products—mainly ash and carbon dioxide—derived from RH gasification. To overcome these challenges, we propose the following:

- ✔ *RH pretreatment for ash removal.* FBG gasification technology is a known technology widely used in coal gasification. However, when agricultural residues are to be used in FBG systems, they must be pre-treated to meet the required specifications and not create operational problems because they contain large amounts of ash. The melting of the ash is a matter of great importance for the successful operation of the FBG systems, because it creates problems of agglomeration of the gasifiers, resulting in the unexpected interruption of the system and financial losses. Solving the ash problem is vital in achieving sustainable bioenergy production [21]. In this study, a combination of a tailor-made pretreatment combining microwave heating with the traditional leaching pre-treatment technique (using water as the solvent) is suggested. With this pretreatment method rice husks-based fuels can be free from the ash constituents. This will result in the decreasing de-fluidization and preventing the operational problems of the SMARt FBG [6].
- ✔ *Alternative uses of gasification ash.* Ash can replace conventional silica sources for making lightweight construction material bricks/blocks [22]. It is suggested that ash could be reused for environmental safety as follows:
  (1) Ash can be used as an insulating material due to its low thermal conductivity.
  (2) Ash can be used as an adsorbent to extract various contaminants from water and air.
- ✔ *NH_3 removal.* Biomass in its various forms often contains nitrogen. NO_x generated from nitrogen bound to RH can cause problems in the gasification system. In that case, it is advised that nitrogen in the form of NH_3 should be removed from the syngas to a minimum [23].
- ✔ *Catalytic tar cracking.* Rice contains ash rich in silicon dioxide, which has a melting temperature, well below the operating temperature range of gasifiers (>800 °C). Rice ash components such as Na, K, Cl, Ca, and Si interact with the bed material to form eutectic mixtures. The melting point decreases, and this creates serious bed

agglomeration problems, especially when silica interacts with silica sand beds (SiO_2) usually used as catalytic material for tar cracking in the reactor [14].

- ✔ *Minimize the liquid gasification waste.* The design of an FBG unit should take care of the potential liquid waste produced in the cooling and purification of syngas unit. These are wastewater and condensate, which require treatment in situ. The condensate is known to contain acetic acid, phenol and many other oxygenated organic compounds that may or may not be soluble in water. This creates a risk for water contamination and side effects on workers' health caused by suspended tar and soluble organic matter. In general, wastewater treatment is usually relatively simple and of low-cost chemical or biological [23]. The most serious component is tar in wastewater. For this an effort should be made to minimize its presence through some operational alterations such as:
  - ✔ To use tar cracking catalytic methods during or after the gasification process (in situ and off site);
  - ✔ To use a hot syngas cleaning method.
  - ✔ To use lower gasification temperatures to reduce the production of tar.

9. Conclusions

Modern agri-food industries face high energy bills and produce large quantities of residues, which could be utilized to provide added value at all levels (material, energy, environmental, economic). Gasification offers an attractive solution allowing the utilization of the waste' energy content to produce energy and fuels to be used on-site or sold to the grid.

A rice husk fluidized bed gasification for a combined heat and power production system of 25,000 t/y capacity enables decentralized energy production from agro-industrial wastes, offering to the agro-industrial sector a circular utilization of resources, and reduction in their environmental footprint. In this study, the assumptions used to simulate the air FBG of rice husks by Aspen Plus software played an important role in the extraction of the results. We assumed that all reactors operate at a constant temperature and the pressure profile and at chemical equilibrium conditions which is not theoretically possible in real reaction conditions. However, in real conditions, the heat loss is higher than the simulated one affecting the whole process energy balance. In addition, for simplification, tar was not considered in the model.

Simulated results show good accuracy in the prediction for H^2 but not for CO, CO_2 and CH_4 content. This indicates that the model needs modification to improve the accuracy of prediction. The results of air gasification showed a deviation from the experimental results varying from −25% to +33%. In general, the deviations in the quantities of the gas components and in the values of LHV, CGE and CCE indicators are not prohibitive. The largest deviations concern the yield of CH_4 and the CGE. The limitations of our model were in assessing tar and char yields, which is a difficult task through a thermodynamic equilibrium model because tar is usually a non-equilibrium product. Since the predictions of mathematical models are substantially improved when tar formation is included, in this study, tar and carbon yields were considered as input parameters and were determined independently of the gasifier operating conditions, according to other bibliographic models. Thus, they have been placed, as inert ingredients, in the R-Gibbs reactor. At high temperatures, such as those examined, the tar content is very moderate, while the gas efficiency is very high. Another reason for the fact that simulated data do not fit very well with the experimental results might be attributed to the fact that the RGIBBS reaction is rather closed to the entrained flow gasifier and not to the fluidized bed gasifier, for which a semi-empirical model might fit better than RGIBBS reaction.

Among the three scenarios examined, the scenario of gasification with steam and oxygen gives a syngas with higher H_2 content resulting in a higher LHV value. However, although, this is a more favorable result, the high thermal requirements of the steam increase the operating cost.

The Aspen Process Economic Analyzer was used for the economic assessment. The ROI and POT were very positive (0.21 and 2.8, respectively), for the case of air gasification with a capacity of 25,000 t/y.

Simulation modeling (SM) and the economic assessment of the planning phase of a gasification system had an increasingly important role in the design of and optimization of the processes and to give an idea of the economic viability of the industrial application, guiding the investors to decide. The main advantages of using SM as a customizable tool to help decision-making is that it makes it possible to analyze how the key process indicators affect the viability of a bioenergy system, without the need to spend more money and time on experimental demonstration.

Author Contributions: Conceptualization, A.Z.; methodology, A.Z.; software, D.A.; validation, D.A.; formal analysis, D.A.; investigation, D.A.; resources, A.Z.; data curation, D.A.; writing—original draft preparation, D.A.; writing—review and editing, A.Z.; visualization, D.A.; supervision, A.Z.; project administration, A.Z. All authors have read and agreed to the published version of the manuscript.

Funding: This research received no external funding.

Institutional Review Board Statement: Not applicable.

Informed Consent Statement: Not applicable.

Data Availability Statement: The data presented in this study are available in https://doi.org/10.3390/su11226433.

Conflicts of Interest: The authors declare no conflict of interest.

Abbreviations

RH	Rice Husk
FBG	Fluidized Bed Gasification
CHP	Combined Heat and Power
ICE	Internal Combustion Engine
EU	European Union
GHG	Greenhouse Gas Emissions
SM	Simulation Modeling
TRL	Technology Readiness Level
ROI	Return of Investment
POT	Pay Out Time
ER	Air Biomass ratio
SBR	Steam Biomass Ratio
LHV	Lower Heating Value
HHV	Higher Heating Value.
CGE	Cold Gas Efficiency
CCE	Carbon Conversion Efficiency
MC	Moisture Content

References

1. Eurorex. 2020. Available online: http://www.eurorex.com (accessed on 4 February 2020).
2. CRES. Available online: http://www.cres.gr/services/istos.chtm?prnbr=24819&locale=el (accessed on 4 February 2020).
3. Adams, P.W.R.; McManus, M.C. Small-scale biomass gasification CHP utilization in industry; Energy and environmental evaluation. *Sustain. Energy Technol. Assess.* **2014**, *6*, 129–140.
4. Gagliano, A.; Nocera, F.; Bruno, M.; Cardillo, G. Development of an equilibrium-based model of gasification of biomass by Aspen Plus. *Energy Procedia* **2017**, *111*, 1010–1019. [CrossRef]
5. Weijuan, L.; Guanyi, C.; Xinli, Z.; Xuetao, W.; Chunmei, L.; Bin, X. Biomass gasification-gas turbine combustion for power generation system model based on ASPEN PLUS. *Sci. Total Environ.* **2018**, *628–629*, 1278–1286.
6. Vaskalis, I.; Skoulou, V.; Stavropoulos, G.; Zabaniotou, A. Towards circular economy solution of rice residues to bioenergy via gasification. *Sustainability* **2019**, *11*, 6433. [CrossRef]
7. Zabaniotou, A.A.; Skoulou, V.K.; Mertzis, D.P.; Koufodimos, G.S.; Samaras, Z.C. Mobile Gasification Units for Sustainable Electricity Production in Rural Areas: The SMARt-CHP Project. *J. Ind. Eng. Chem. Res.* **2010**, *50*, 602–608. [CrossRef]

8. Tadeu, A.; Marques, B.; Almeida, J.A.S.; Pinto, V. Application of rice husk in the development of new composite boards. Construction and building materials: A review. *J. Asian Ceram. Soc.* **2018**, *6*, 299–313.
9. Sarangi, M.; Bhattacharyya, S.; Behera, R.C. Effect of temperature on morphology and phase transformations of Nano crystalline silica obtained from rice husk. *Phase Transit.* **2009**, *82*, 377–386. [CrossRef]
10. Della, V.P.; Kuhn, I.; Hotza, D. Rice husk ash as an alternate source for active silica production. *Mater. Lett.* **2002**, *57*, 818. [CrossRef]
11. Kishore, R.; Bhikshma, V.; Prakash, P. Study on strength characteristics of high strength rice husk ash concrete. *Procedia Eng.* **2011**, *14*, 2666–2672. [CrossRef]
12. Phyllis2 Data Base ECN—Rice Husk Agrino Industry Greece. Available online: https://plyllis.nl/Browse/Standard/ECN-Phyllis#rice%20husk (accessed on 7 September 2019).
13. Juan, M.; Michel, V.; Matteo, B.; Encarnación, R. Modeling and model performance evaluation of sewage sludge gasification in fluidized-bed gasifiers using Aspen Plus. *J. Air Waste Manag. Assoc.* **2019**, *69*, 23–33.
14. Damartzis, T.; Michailos, S.; Zabaniotou, A. Energetic assessment of a combined heat and power integrated biomass gasification–internal combustion engine system by using Aspen Plus®. *Fuel Process. Technol.* **2012**, *95*, 37–44. [CrossRef]
15. Mansaray, K.G.; Al-Taweel, A.M.; Ghaly, A.E.; Hamdullahpur, F.; Ugursal, V.I. Mathematical Modeling of a Fluidized Bed Rice Husk Gasifier: Part I-Model Development. *Energy Sources* **2010**, *22*, 83–98.
16. Lingqin, L.; Yaji, H.; Changqi, L. Prediction of Rice Husk Gasification on Fluidized Bed Gasifier Based on Aspen Plus. *Bioresources* **2011**, *11*, 2744–2755.
17. Tungalag, A.; Lee, B.; Yadav, M.; Akande, O. Yield prediction of MSW gasification including minor species through ASPEN plus simulation. *Energys* **2020**, *198*, 117296. [CrossRef]
18. Tamer, M.M.; Monteiro, E.; Ramos, A.; El-Salam, M.A.; Abel, R. An Eulerian model for forest residues gasification in a plasma gasifier. *Energy* **2019**, *182*, 1069–1083.
19. Peters, M.; Timmerhaus, K.; West, R. *Plant Design and Economics for Chemical Engineers*, 5th ed.; McGraw-Hill: New York, NY, USA, 2006; ISBN 13-978-0072392661.
20. Eurostat. 2020. Available online: https://ec.europa.eu/eurostat/web/europe-2020-indicators (accessed on 25 March 2020).
21. Chakraborty, M. Beneficial Reuse of Rice Husk. EM Magazine, a Copyrighted Publication of the Air & Waste Management Association (A&WMA). January 2020. Available online: www.awma.org (accessed on 10 April 2020).
22. Aziz, M.; Darmawan, A.; Fotriano, A.C.; Tokimatsu, K. Integrated system of rice husk in the development of new composite boards. *Constr. Build. Mater.* **2018**, *176*, 432–439.
23. Bridwater, A.V.; Kaltschmitt, M. Biomass Gasification and Pyrolysis. In Proceedings of the Biomass Gasification Network and Biomass Pyrolysis Network (PyNE), Stuttgart, Germany, 9–11 April 1997; CPL Press: Newbury, UK, 1997. ISBN 1-872691-71-4.

Article

Are Short Food Supply Chains More Environmentally Sustainable than Long Chains? A Life Cycle Assessment (LCA) of the Eco-Efficiency of Food Chains in Selected EU Countries

Edward Majewski [1], Anna Komerska [2], Jerzy Kwiatkowski [2], Agata Malak-Rawlikowska [1,*], Adam Wąs [1], Piotr Sulewski [1], Marlena Gołaś [1], Kinga Pogodzińska [1], Jean-Loup Lecoeur [3], Barbara Tocco [4], Áron Török [5], Michele Donati [6] and Gunnar Vittersø [7]

[1] Institute of Economics and Finance, Warsaw University of Life Sciences—SGGW, 02-787 Warsaw, Poland; edward_majewski@sggw.edu.pl (E.M.); adam_was@sggw.edu.pl (A.W.); piotr_sulewski@sggw.edu.pl (P.S.); marlena_golas@sggw.edu.pl (M.G.); kinga.pogodzinska@wp.pl (K.P.)

[2] Faculty of Building Services, Hydro and Environmental Engineering, Warsaw University of Technology, 00-653 Warsaw, Poland; anna.komerska@pw.edu.pl (A.K.); jerzy.kwiatkowski@pw.edu.pl (J.K.)

[3] CESAER, AgroSup Dijon, INRA, University Bourgogne Franche-Comté, F-21000 Dijon, France; jeanloup.lecoeur@outlook.fr

[4] Newcastle University Business School, Newcastle University, Newcastle upon Tyne NE1 4SE, UK; Barbara.Tocco@newcastle.ac.uk

[5] Department of Agricultural Economics and Rural Development, Corvinus University of Budapest, 1093 Budapest, Hungary; aron.torok@uni-corvinus.hu

[6] Department of Economics and Management, Università degli Studi di Parma, 43125 Parma, Italy; michele.donati@unipr.it

[7] SIFO Consumption Research Norway, Oslo Metropolitan University, 0130 Oslo, Norway; gunjar@oslomet.no

* Correspondence: agata_malak_rawlikowska@sggw.edu.pl; Tel.: +48-225-934-220

Received: 17 August 2020; Accepted: 11 September 2020; Published: 16 September 2020

Abstract: Improving the eco-efficiency of food systems is one of the major global challenges faced by the modern world. Short food supply chains (SFSCs) are commonly regarded to be less harmful to the environment, among various reasons, due to their organizational distribution and thus the shortened physical distance between primary producers and final consumers. In this paper, we empirically test this hypothesis, by assessing and comparing the environmental impacts of short and long food supply chains. Based on the Life Cycle Assessment (LCA) approach, we calculate eco-efficiency indicators for nine types of food distribution chains. The analysis is performed on a sample of 428 short and long food supply chains from six European countries. Our results indicate that, on average, long food supply chains may generate less negative environmental impacts than short chains (in terms of fossil fuel energy consumption, pollution, and GHG emissions) per kg of a given product. The values of eco-efficiency indicators display a large variability across analyzed chains, and especially across different types of SFSCs. The analysis shows that the environmental impacts of the food distribution process are not only determined by the geographical distance between producer and consumer, but depend on numerous factors, including the supply chain infrastructure.

Keywords: eco-efficiency; environmental impact; GHG emissions; energy consumption; food chain; short food supply chains (SFSCs); Life Cycle Assessment (LCA)

1. Introduction

The environmental impacts of food distribution systems are primarily related to the transportation of goods. Improving the eco-efficiency in transport, currently estimated at around 15% of global greenhouse gas (GHG) emissions [1], represents one of the major challenges of the modern World [2–4], becoming one of the goals of the EU climate and energy policy [5].

The shortening of supply chains, which is usually associated with shortening the distance over which agri-food products "travel", is often treated as a strategy to reduce energy consumption in transport and improve environmental performance of distribution processes in general. This is quite commonly considered a factor in reducing transportation-related negative externalities [6–8], but this stereotypical view is being more and more frequently questioned [9–12].

In this context, the promotion of short food supply chains (SFSCs) to support environmental sustainability is an increasingly debated topic [7,8,13–15]. The European Rural Development Regulation (1305/2013) defines a short supply chain as one that "has a limited number of economic operators, committed to co-operation, local economic development, and close geographical and social relations between producers, processors and consumers" [16]. The specific feature of SFSCs, referring to this definition, is the existence of a limited number (usually maximum one) of intermediaries between the farmer (primary producer) and the final consumer. The concept of short food supply chains has been particularly promoted in the EU in recent years, constituting an important element of the Rural Development Policy [16]. This is because of envisioned advantages of short food distribution models, which include, among others, economic value added for food producers and social benefits, such as the identification of a food's place of origin and direct contact with the producer, which are valued highly by many consumers [17]. It is also believed that short food supply chains generate environmental benefits by shortening the distance travelled by food (agricultural products). This is, however, an assumption that requires challenging and which will be empirically discussed in this paper.

Some researchers note that due to the small share of transport in total GHG emissions generated in food production, distribution and consumption, the issue of food supply chain length is of little importance for environmental sustainability [11,18,19]. However, the 9.7% share of logistics processes in the estimated total energy consumption of production and distribution [20] cannot be considered unimportant and, thus, the environmental impacts of distribution models, including transport, must not be neglected [20–24].

Food supply chain management is a complex process, largely because of specific requirements related to functional characteristics of food and fresh food in particular [25–27]. It requires, among others, maintaining the continuity of supply that is the basis of food security [28–31], which induces systematic and continuous transport activities. The logistics infrastructure (the assets of the wholesalers and retailers, including cooling equipment) plays an important role in the food distribution process, which is necessary to ensure proper conditions for food storage and delivery to the final consumer.

To date, most of the environmental impact assessments of supply chains presented in the literature are qualitative and limited to Food Miles calculation and/or the transportation-related emissions of greenhouse gases, expressed in CO_2 equivalent. However, there is a lack of comprehensive analysis of environmental impacts of food supply chains of different organizational structures and lengths, which would cover a whole complexity of distribution chains, while taking into account the life cycle of all assets employed.

In our paper, we attempt to fill in this existing gap. The main goal of the article is to empirically assess the environmental impacts of short and long food supply chains through the estimation of selected eco-efficiency indicators. Referring to the views commonly presented in the literature on the environmental benefits of short food supply chains (SFSCs) we attempt to verify the main hypothesis stating that short food supply chains are less harmful to the environment than conventional, long supply chains.

The methodology applied in our study is based on the Life Cycle Assessment (LCA) approach and the eco-efficiency concept.

The paper is structured as follows: the introductory section defines the research problem and presents the objective of the study. Section 2 provides a discussion of the main literature on short and long food supply chains and associated environmental impacts and introduces the conceptual framework. This is followed by Section 3, which explains the research methodology. Section 4 provides a discussion of the main research findings and Section 5 summarizes the key conclusions from the study.

2. Literature Review

2.1. Short vs. Long Food Supply Chains and Their Environmental Impact

Historically, there were two basic food supply models for food market-direct deliveries to consumers and sales on farmer's markets. Their importance was systematically diminished mainly due to the growth of the food processing sector. As a result, as well as due to processes of industrialization, urbanization, and the development of long-distance transportation [32], the supply chains were extended to include intermediaries and expand transportation functions. In the course of these changes, the market became dominated by logistically complex mass distribution systems; although, compared to many highly globalized supply chains, the food market is less concentrated. Nowadays, a renaissance of traditional forms of food sales can be observed, but other, innovative types of short supply chains are also emerging [12,33].

The distinguishing feature of short supply chains is a specific type of relation between key actors of the food chain—primary producers (basically farmers)—and consumers [34,35] that includes "face to face" contact (purchases directly from producers), "spatial proximity" (production and distribution locally) and "spatial extension" (consumers have information about the place and production process).

Referring to this type of relation, an approach to defining short supply chains was proposed by Malak-Rawlikowska et al. [12] based on the three "proximity" dimensions:

- "geographical proximity", which expresses the physical distance of transportation, measured with Food Miles, travelled by the product from the location of production to the living place of the final consumer;
- "social proximity", which refers to the close 'relationship' between the producer and consumer of the food, resulting in the direct transfer of information and mutual trust;
- "organizational proximity", which is related to the number of intermediaries in the food supply chain.

It is believed that shortening the distance that food products travel before reaching the consumer, which is one of the fundamental characteristics in the concept of short food supply chains [7,8,13–15,36], should reduce the negative environmental externalities related to transport (energy consumption, GHGs and other emissions) [11,15,18,19,37]. However, Kneafsey et al. [15] (p. 32) underline that several publications characterized SFSCs as "beneficial for the environment", however without providing "any further qualitative or quantitative evidence to substantiate claims made". Moreover, Galli et al. [10] (p. 9) emphasize that "SFSCs are not by definition more environmentally friendly than conventional, longer supply chains". However, it is proven that SFSCs can bring economic benefits to farmers and the local economy, as well as to increase consumer confidence in the products they buy and their producers [6,8,10,14,15,19,38].

Economic and social benefits (close co-operation, geographical and social relations between producers, processors and consumers) that may be attributed to short food supply chains appear to be indisputable. There is growing evidence that short distribution chains indeed provide added value for producers [8,10,12,15,19,38,39] and generate social benefits [15,17,40–42].

Commonly shared opinions on the environmental benefits of using short food supply chains are based on mainly qualitative assessments, shaped by the simple association of geographical proximity and short transportation distances with low energy expenditure and relatively less harmful environmental impacts. This opinion, however, is hardly defendable in light of the most recent studies.

Gonçalves and Zeroual [43] and Mancini et al. [44] emphasize that the need for frequent, multiple deliveries of small quantities of products may have negative impacts on environmental sustainability; however, according to Bloemhof and Soysal [45], there are negative impacts, but they are "not so important". The results of quantitative assessments of sustainability of short food supply chains from the study conducted within the Strength2Food (The European Union's Horizon 2020 research and innovation project "Strengthening European Food Chain Sustainability by Quality and Procurement Policy" (grant agreement no. 678024)) project, although limited to transportation activities, prove that doubts concerning environmental benefits of SFSCs are fully justifiable [12].

The burden on the environment with the negative consequences of food distribution processes is determined not only by the geographical distance between the farmer (producer) and the consumer. There are several factors to be taken into account, such as the infrastructure along the supply chain, the type (capacity) of vehicles used, as well as the conditions in which food is transported [21,46], stored (e.g., refrigerators, freezers, etc.) and displayed in retail outlets.

The need for economical management of natural resources and the minimization of negative externalities in line with the Sustainable Development paradigm is of particular importance in the context of accelerating climate change. This increases the pressure to search for solutions that reduce energy consumption and GHG emissions not only in the sphere of food production, but also in general in distribution processes [21,47–51].

2.2. Eco-Efficiency Concept and Life Cycle Assessment to Measure Environmental Impacts

The concept of eco-efficiency, which appeared in the literature in the early 1990s, was disseminated widely by the World Business Council for Sustainable Development (WBCSD) as an approach to strengthen the competitiveness and environmental responsibility of enterprises [52–54]. Nowadays, the concept of eco-efficiency is perceived as a quantitative tool for the simultaneous assessment of economic and environmental aspects of economic systems [55], being considered one of the most important instruments used in sustainability assessments [56–58]. The practical implementation of the eco-efficiency approach leads to the delivery of "competitively priced goods and services that satisfy human needs and bring quality of life while progressively reducing environmental impacts of goods and resource intensity throughout the entire life cycle to a level at least in line with the Earth's estimated carrying capacity" [59]. Maxime et al. [60] consider the eco-efficiency approach an effective way to evaluate Sustainable Development parameters aimed at reducing the consumption of natural resources and mitigation of negative environmental impacts of manufacturing processes.

Eco-efficiency can be defined as the relation of an effect, most often expressed as the value of products at the firm, sector or even the entire economy level, and inputs constituting a measure of environmental pressure generated by this firm, sector or the economy [61]. Gómez-Limón et al. [53] emphasize that eco-efficiency reflects the possibility of achieving specific economic results with the minimal use of natural resources, causing the least possible damage to the environment.

In the most general approach, the measurement of eco-efficiency is based on partial or composite indicators reflecting the relationship between specific environmental categories, which express the impacts of the production system on selected elements of the natural environment, and economic effects that reflect the production and economic performance of this system. In our study, for the eco-efficiency measurement, several eco-efficiency indicators were used, as derived from the general equation [62]:

$$\text{ECO}-\text{EFFICIENCY}=\frac{\text{ENVIRONMENTAL COSTS (INPUTS)}}{\text{ECONOMIC EFFECTS (BENEFITS)}}$$

The Life Cycle Assessment methodology offers the most comprehensive, but also the most complex, approach to the environmental impact analysis, thus providing the possibility of estimating a variety of specific eco-efficiency indicators [63].

The origin of the use of LCA for eco-efficiency measurements is related to the methodology proposed in 1996 by BASF [64]. The originally proposed methodology was aimed at supporting business projects by enabling the assessment of various production solutions while taking into account their long-term environmental impacts and their economic significance.

Concepts of eco-efficiency measurement and Life Cycle Assessment have been primarily applied in analyses related to industrial products and processes, also covering transportation. In recent years, there have been a growing number of applications related to the agri-food sector that focus on farm-level assessments of production systems [55,65–68].

LCA-based eco-efficiency assessments of entire food supply chains are limited so far and are related mainly to single products. Usually, these analyses do not focus on the distribution chains themselves, but on the entire life cycle of agricultural and food products, covering agricultural production and the processing phase [11,18,19,69,70].

To the best of our knowledge, however, the concept of the LCA methodology to assess eco-efficiency of various types of food distribution chains, defined according to the proximity dimensions, has not yet been applied.

In presenting the results of our study, we intend to suggest a methodological approach that may be best suited to this type of analysis.

A comprehensive assessment of the environmental impacts of various supply chains requires considering all effects generated throughout the product or system's life cycle [71], including energy consumption as well as the use of capital assets involved. Due to its holistic nature, Life Cycle Assessment is a universal approach to quantitative environmental analyses. Despite some criticism and limitations [72,73], the LCA methodology is a commonly applied and recognized method for environmental assessments brought into elementary level, referring to a single material and product. The environmental performance of a product is generally shown in an Environmental Product Declaration (EPD) where environmental indicators are defined and standardized according to a set of standards on LCA (ISO 14040, ISO 14044, ISO 14025 and EN15804) or in the Environmental Certificate provided by the manufacturer. In recent years, the EPD database has grown substantially and it now covers a wide range of categories, including services, materials, vehicles, packages, food, machinery, chemical products and many other. More information on the environmental impacts of various products or materials can also be found in the literature. Regarding transportation, the share of transport in global greenhouse gases (GHG) in typical modes of transport [74–77] as well as electric vehicles [76,78,79] was analyzed in a search for ways of reducing emissions and global energy consumption [1]. Despite a considerable amount of literature studies concerning the environmental impact of different transportation modes [76,80–82], most of the studies are focused only on the level of greenhouse gas emissions expressed in the Global Warming Potential (GWP) index.

Regarding the construction industry, LCA is used mainly to assess the environmental performance of buildings through their life stages, including emissions from building materials, the construction process, the use of energy and demolition [83–86]. Therefore, it is often applied to building certification systems and building benchmarking [87–90] or used as a tool in decision processes towards sustainable building or neighborhood design [91–93]. Many studies show that, in conventional buildings, the majority of the energy and GHG emissions, accounting for 80–90% of the total carbon footprint, are related to the operational stage [94,95].

There are examples of using the LCA approach in the evaluation of supply chain configurations [96,97]. The LCA methodology is also applied to agricultural production and food products. Taking this perspective may lead to conclusions about a relatively small share of transport and distribution in the total loads generated in the life cycle of agri-food products, resulting possibly in recommendations to search for ways to reduce emissions other than via distribution links in the food supply and consumption system [11,18,19]. Similarly, in other industries, the inclusion of all life cycle phases in the LCA assessment makes the production phase the dominant factor. On the other hand, narrowing the scope of the analysis to the distribution segment may suggest that the distance over which products are transported plays a more important role [77]. The case is not clear-cut, however, as it can also be pointed out that in some cases the transportation link is responsible for even 50–70% of the total carbon footprint, e.g., when fresh vegetables or fruits are transported by energy-intensive means of transport over considerable distances [18]. The assessment of the environmental impacts generated beyond farm gate should not, however, be limited only to food transportation related to emissions from fuel combustion. It should include the whole complexity of food distribution chains such as the necessary infrastructure of all kinds of stores, the energy needed for keeping stores running, cooling energy for food preservation and emissions from the production of transportation vehicles. The LCA is well suited to make such complex assessments.

3. Materials and Methods

3.1. Primary Data Collection

The eco-efficiency LCA assessment was performed on a sample of 191 food producers (169 farmers and 22 fishmongers) participating in short and long food supply chains in six European countries: France, Hungary, Italy, Norway, Poland and the United Kingdom (Table 1). Fishmongers were integrated in the United Kingdom and Norway samples due to the importance of the fish industry in the agri-food sector in these countries. It should be emphasized here that fishmongers' activities are not agricultural or land based (as in the case of fish farms), but have been investigated as types of market channels used by farmers. In order to simplify the analysis in the paper, we will be referring to 'farms' from now on.

Table 1. Number of farms and chains in the research sample.

Country	France	Hungary	Italy	Norway	Poland	United Kingdom	Total
Number of farms	22	39	22	16	57	35	191
of which:							
Farms	22	39	22	14	57	15	169
Fishmongers	-	-	-	2	-	20	22
Number of chains	65	79	63	32	120	69	428
of which:							
short chains	52	61	53	24	54	56	300
long chains	13	18	10	8	66	13	128

Source: own elaboration.

The total number of chains detected in the sample noticeably exceeds the number of farms, because individual farmers use several distribution paths to sell the same product [12] (see Section 3.2). Sales to processors are not covered by the present study, because the analysis of the distribution channels downstream of food processing is beyond the scope of our research.

The selection of farms included in the sample meets the following two criteria: (1) farms participating in at least one type of SFSC; (2) farms operating in one of the product categories defined in Table 2. According to these criteria, selected farms were interviewed in relation to the food products belonging to the specified product categories, so that the analysis of the farms' distribution channels related only to these products.

Table 2. Number of chains used by producers for distribution of products in the sample.

Category of Product	Total Number of Chains	Short Chains	Long Chains
Fruits	96	46	50
Vegetables	89	77	12
Fish and Seafood	46	36	10
Cheese	93	66	27
Meat	45	39	6
Honey	32	23	9
Eggs	6	5	1
Total	428	300	128

Source: own elaboration.

The sample is not representative of the whole population of food distribution chains and farms across countries. A representative sample would require random sampling, which is not feasible, since databases with information on the structure of distribution channels is not available. Nevertheless, this large sample allows for a detailed comparison of short and long food supply chains and for drawing conclusions on the regularities observed in food distribution.

Data were collected through surveys conducted between November 2017 and November 2018 with the use of a dedicated questionnaire. All the questions were tested in pilot surveys conducted in Poland and in France. The questionnaire covered the following areas: farm description (labor, production structure, means of transportation, turnover); sales (quantities sold to different distribution chains, prices, locations and distances to final destinations); specific distribution related data (amounts transported in single deliveries, labor inputs, costs of packaging, other distribution costs); self-assessment of bargaining power and chain evaluation by producers.

3.2. Typology of Food Supply Chains

There are several types of food supply chains that may be distinguished in relation to different parameters such as the product's destination, the number and roles of intermediaries involved, and the type of product [15,34,35,98–100]. In accordance with the study by Malak-Rawlikowska et al. [12], nine 'short' and 'long' distribution chains were selected for our study (Table 3). All chains with none or a single intermediary between consumer and producer were categorized as 'short', whereas supply chains with more than one intermediary were considered 'long'.

Table 3. Participation of producers, intermediaries and consumers in transportation activities in selected food supply chains.

Chain	Producer Gate	Pink = Consumer Travel (R2C); Blue = Product Travel (P2R)				
		Short chains				
a. Pick-your-own	Producer					Consumer
b. On-farm sales to consumers	Producer					Consumer
c. Internet sales—courier deliveries	Producer/Courier					Consumer
d. Direct deliveries to consumer	Producer					Consumer
e. Sales on farmers' markets	Producer				Farmers' Market	Consumer
f. Direct deliveries to retail	Producer				Retail Shop **	Consumer
		Long chains				
g. On-farm sales to intermediaries	Producer	Agent	Wholesaler ***		Retail Shop **	Consumer
h. Sales on wholesale market	Producer		Wholesaler		Retail Shop **	Consumer
i. Sales to hypermarket chains	Producer	Producers' Group **		Logistics Centre	Hypermarket Store	Consumer

* The pink color in the table indicates a part of the physical distance in the distribution channel in which food is transported from the purchase (sales) by the consumer. Blue color indicates that the product travels from the farm gate to sales point being transported by producers or intermediaries. ** Retail outlets including food stores, hotels and restaurants. *** Alternatively. Source: Malak-Rawlikowska et al. 2019 [12].

Our assessments were limited to the main stages of the distribution chain, which we broadly define as a system of moving products from the farm (producer) gate to the end consumer. Production (farming) systems were not a subject for the analyses, assuming the same technologies of production were used, irrespective of the future decisions on the choice of distribution channel.

In the eco-efficiency assessments, two types of transportation activities within the distribution chain were taken into account, namely:

- transporting products from the primary producer (farm gate) to a retail outlet (P2R)—performed by producer or intermediary (e.g., agent, wholesaler, producers' group, logistic center of the hypermarket chain),
- transporting food by consumer from retail outlet to consumers' place of living (R2C).

The participation of specific actors in the distribution process is illustrated in a graphic form in Table 3.

Transportation activities in short channels (*a–f*) are split between suppliers and consumers in different proportions:

- In *f. Direct deliveries to retail* (the only short food supply chain containing an intermediary ('Retail Shop') between producer and food consumer), primary producers and different intermediaries deliver goods to retail outlets, from where the food is transported by consumers;
- Transportation in chains *a. Pick-your-own* and *b. On-farm sales to consumers* is done entirely by consumers.
- In channels *c. Internet sales—courier deliveries* and *d. Direct deliveries to consumer* producers deliver products to consumers, by themselves (*d*) or by out-sourcing transportation to courier companies (*c*). Couriers are not treated as intermediaries, but as providers of transportation services only.
- In chains *e. Sales on farmers' markets* and *f. Direct deliveries to retail* producers transport products to retailing places (Farmers' Market or Retail Shop), from where the food is transported by consumers.

Each of the long chains contains at least two intermediaries:

h. Sales on wholesale market: an agent purchasing products for re-sale, possibly also a wholesaler, for sale by a retailer;
g. On-farm sales to intermediaries: wholesaler and retailer;
i. Sales to hypermarket chains: logistic center, possibly a producers' group, hypermarket.

All of the above types of chains were represented in the sample. An individual producer in the sample used two and four chains on average, within the range from two (minimum) to five (maximum), both short and long. About 84% of producers used at least one short chain, and 51% participated in at least in one long chain (Table 4).

Although SFSCs represented a large majority of the sample, accounting for 70% of the total 428 chains used by farmers (Table 4), only 36.6% of the total quantity sold was delivered through short chains. This is because SFSCs typically suffer from diseconomies of scale (especially considering small to medium scale producers), and cannot absorb large quantities of produce due to limited consumer demand, as well as limited farm labor resources to dedicated to distribution and retail. These restrictions lead farmers to diversify their distribution chains.

Table 4. Yearly sales by supply chain for the full sample.

Supply Chains	Total Volume Sold and Market Share		Producer Participation Across Chains		
	(Tons)	(%)	Number of Farms Using Certain Type of Chains	As % of All Chains Used n = 428	As % of All Farms in the Sample n = 191 *
a. Pick-your-own	16.3	0.1	3	0.7	1.6
b. On-farm sales to consumers	854.7	7.1	110	25.7	57.6
c. Internet sales—courier deliveries	148.2	1.2	28	6.5	14.7
d. Direct deliveries to consumer	176.7	1.5	28	6.5	14.7
e. Sales on farmers' markets	313.1	2.6	73	17.1	38.2
f. Direct deliveries to retail	2872.7	24.0	58	13.6	30.4
Short Chains Total	4381.7	36.6	300	70.1	84.3
g. On-farm sales to intermediaries	2266.3	18.9	39	9.1	20.4
h. sales on wholesale market	2315.1	19.3	60	14.0	31.4
i. Sales to hypermarket chains	3018.9	25.2	29	6.8	15.2
Long Chains Total	7600.3	63.4	128	29.9	51.3
Total sample	11,982.0	100	428	100	-

* note that one farm can use diverse chains at the same time therefore this column should not be summarized to 100%. Source: own elaboration.

3.3. *Environmental Assessment of Food Chains*

The environmental impacts of food distribution chains were evaluated using the Life Cycle Assessment methodology with respect to a set of standards concerning LCA approach. The four LCA phases—goal and scope definitions, life cycle inventory, life cycle impact assessment and interpretation of the outcomes—were defined and performed according to ISO 14040 [101] and ISO 14044 [102]. Life Cycle Assessment was performed using the OneClick LCA calculation tool [103] which is an engineering software used for life cycle studies. It complies with the international standards, requirements and building certification schemes such as LEED and BREEAM.

3.3.1. Goal and Scope Definition of Environmental Assessment

Aim of the LCA Analysis

Short and long chains of food distribution involve various assets (buildings, equipment, means of transportation). The assessment of their influence on the environment is based on six most commonly used and recognized environmental impact categories: Global Warming Potential (GWP), Acidification Potential (AP), Eutrophication Potential (EP), Ozone Depletion Potential (ODP), Photochemical Ozone Creation Potential (POCP) and Non-Hazardous Waste Disposed. If related to the production volumes, they form a set of eco-efficiency indicators. All above mentioned environmental indicators were normalized per 1 kg of produce sold in each of the analyzed distribution chains. This is the functional unit, that is, the reference by which the input and output flows of material and energy along the distribution chain are retrieved.

The goal and the scope of the analysis are summarized in Table 5.

Table 5. Summary of scope/definitions.

	Life Cycle Assessment of Food Supply Chains	
Functional Equivalent	Food supply chain	Short and long food distribution chains
	Definition of supply chains	From the farm gate to the consumer
	Location	European Union
	Functional unit	1 kg of purchased product
	Assessed impact categories	GWP, AP, EP, OPD, POCP, Non-Hazardous Waste Disposed
	Eco-efficiency indicator	Assessed impact category/1kg of product
LCA System Boundary	Cradle to grave	
	Buildings: Building construction materials, including technical equipment and furnishings Energy in operation stage, excluding office equipment Emissions due to refrigerant leakage included	
	Transport vehicles: Manufacturing, fuel use and disposal included Road infrastructure excluded Emissions due to refrigerant leakage from air conditioning excluded	
Calculation Software	OneClick LCA© and 360optimi, Bionova [103]	

Source: own elaboration.

LCA System Boundaries

The system boundary for the food supply chain is defined in our study as being from the farm gate to the consumer. Depending on the organizational and geographical proximity of chains, as well as on chain infrastructure and types/means of transportation used, the emission level can substantially differ. Therefore, the assessment includes the emission level from buildings and transport vehicles using a 'cradle-to-grave' approach, i.e., from the extraction of raw materials, material processing and manufacture, through to the use phase and, finally, to the end of the life phase, which includes disposal and material recycling.

Figure 1 identifies the phases embedded in the system boundary of buildings according to the European standard EN 15978 [104]. Stages B2-B3, related to maintenance and repair services, were excluded from the analysis due to lack of sufficient data; however, they represent a negligible environmental effect compared to the remaining life cycle stages.

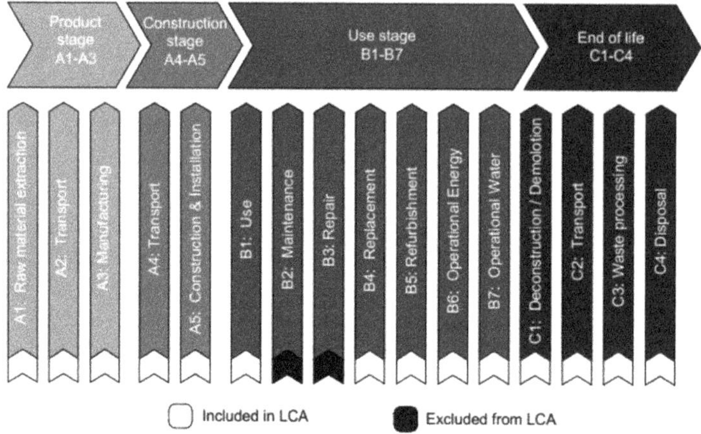

Figure 1. Life Cycle Assessment (LCA) system boundary definition applied to buildings (according to EN 15978).

Regarding transportation vehicles, LCA system boundaries were defined according to the Environmental Product Declaration (EPD), developed in line with Product Category Rules (PCR) and the principles described in ISO 14025 [105]. The life cycle of vehicles can be divided into 3 main life stages: manufacturing, use phase and end of life. The system boundary applied to transportation vehicles is shown in Figure 2.

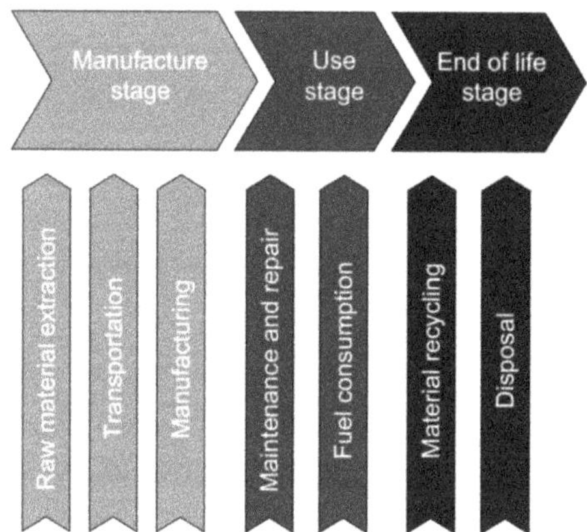

Figure 2. LCA system boundary definition applied to transportation vehicles.

3.3.2. Life Cycle Inventory

Buildings

Different types of buildings with appropriate infrastructures were assigned to food distribution chains: small store, medium store, large store, wholesale market and farmers' market instalments.

The types of trading outlets are consistent with those used in national statistics [106]. For the purpose of this study, the average size (commercial area), average sales volume, and main construction characteristics for each type of trading outlet were defined as shown in Table 6.

Table 6. Definition of trading types.

Type of Trading Outlet	Size Category (m^2)	Sales Area (m^2)	Parking Area (m^2)	Average Sales Volume (kg/m^2 year)	Construction Type	Infrastructure Type (Parking)
Small store	<399	100	25	1208	Typical concrete structure, brick wall insulated	Asphalt paving
Medium store	400–2499	1000	250	1116	Lightweight construction: steel frame with metal cladding, insulated	Asphalt paving
Large store (Hypermarket type)	>2500	5000	1250	1018	Lightweight construction: steel frame with metal cladding, insulated	Asphalt paving
Wholesale market	X	80,300	39,700	4480	Lightweight construction: steel frame with metal cladding, non-insulated, Paved area with steel roof	Asphalt paving
Farmer's market	X	6000	500	130	Paved area with steel roof	Concrete pavement tiles

Source: own elaboration.

Considering the high variability in the size and equipment used in different types of retail stores, simplifications in making relevant assumptions were unavoidable. It should be emphasized, however, that these simplifications should not affect the calculation of eco-efficiency indicators since volume of sales per unit of sales area, which is an essential parameter in the assessments, does not differ significantly within the typology of trading outlets.

Two types of buildings were excluded from the analysis: logistics centers of hypermarket chains and transfer facilities of courier companies involved in the Internet sales chain. It was decided that in both cases, the fresh foods under consideration constitute an extremely low share of the turnover and the length of their stay in storage is minimal. Hence, their participation in generating environmental impacts in the LCA analysis may be considered negligible.

Several assumptions were made regarding the construction of buildings and infrastructure. More specifically, it was assumed that:

- a small store is a typical building with a concrete structure, with walls made of concrete blocks insulated with extruded polystyrene and few windows;
- small and large stores are defined as single-story halls characterized by a steel lightweight frame, with walls made of sandwich panels with a mineral wool core and double steel siding, a metal roof insulated with mineral wool and an industrial floor;
- wholesale markets are represented by two types of trading buildings: a steel frame hall with non-insulated walls and a paved area with a steel roof;
- farmers' markets consist usually of an outdoor site where famers sell their products either directly from their cars or tables and stands hosted in a paved area with concrete paving and paved area with steel roofing.

For each type of store, a separate life cycle inventory (LCI) was created, involving building materials and energy for building construction, energy consumption in the use phase and the amount of refrigerant, including its average annual leakage. For every store type, the same reference building lifespan of 40 years was adopted. The assumptions related to buildings and infrastructure applied to the LCA study are summarized in Table 7.

Table 7. Summary of Life Cycle Assessment analysis assumptions for buildings.

Area of Analysis	Data Sources and Assumptions
Material quantities (A1–A3)	Data inventory based on real reference case buildings and technical buildings' documentation Material database: Ecoinvent, GaBi Environment data source: Environmental Product Declaration (EPD) Average regional European data if possible. Otherwise, local manufacturer was selected
Building material transport distances (A4)	Average European transport distance specific for each material type
Construction and installation process (A5)	Average emissions for the construction process were based a general scenario available in the calculation tool
Material service lifetime (B4–B5)	According to the Environmental Product Declaration (EPD)
Building use phase energy consumption (B6)	Energy consumption was based on the real measured data of different stores. Emissions from grid electricity were calculated according to the European energy mix
End-of-Life Stage (C1–C4)	Based on a scenario provided in the calculation tool
Total lifetime TLT	40 years

Source: own elaboration.

Material Production Phase (A1–A3)

Material types and quantities were estimated according to the technical building documentation of real trading outlets. All building flows of materials and energy were included in LCI and attributed to the building element category: external walls (envelope, structure and finishes), external windows and roof lights, foundations, structural frame, roof (including coverings), floors, walls and partitions, internal doors, internal floor finishes, internal ceiling finishes, internal wall finishes. Additionally, the LCI covered equipment such as a cold store within outlets, a dry cooler and industrial shelves. The specific type of material or product was selected from a database provided by the LCA software. Average European data were chosen. In the event that no average data were available, a local manufacturer was selected for filling in the missing information. The estimation of the energy consumption impact refers to the European energy mix.

Construction Phase (A4–A5)

The level of emissions generated by machines, vehicles and by heating/cooling systems during the construction stage was calculated in accordance with the general scenario provided by the LCA software. European transport distances from the manufacturer to construction site, typical for a specific type of material, were applied with one of the following types of transportation:

- An 8-m^3 concrete mixer truck, 100% fill rate: used for the transportation of a concrete mixer.
- A 40-ton capacity truck, 100% fill rate: used for the transportation of large-scale materials, such as steel, windows, concrete elements, insulation, etc.
- A 19-ton dumper truck, 100% fill rate: used to transport loose materials, such as sand, soil substrates, gravel.

Use Phase (B1, B4–B7)

During a building's life cycle, considerable quantities of energy are used for heating, cooling, lighting and powering equipment. It was assumed that all types of outlets were using R404A for food refrigeration storage and air conditioning. Emissions related to refrigerant leakage were estimated using the literature data [107,108] and the yearly loss rates for the cooling systems were assumed as follows: 10% for small shops, 15% for medium shops and 20% for large stores. Since the amount of refrigerant depends on the cooling capacity and cooling volume, the average consumption of refrigerant was assumed to be equal to 0.3 kg/m^2 for small shops, 0.2 kg/m^2 for medium shops and 0.1 kg/m^2 for large stores.

Materials and building components are characterized by a building's service lifetime, therefore, some of them will be replaced during the assumed building life cycle. Material service lifetime was defined according to the Environmental Product Declaration (EPD), if available. Otherwise, default values of the material replacement periods were applied in the analysis according to the average data provided in the LCA tool.

The quantity of energy required depends on the building type and equipment elements (i.e., cooling counters in the case of stores, or refrigerated storage rooms in storage units). Therefore, for the purpose of this analysis, the data on heat and electricity consumption were obtained through interviews at a sample of retail outlets. The energy use indicators related to the building area for each type of outlet are listed in Table 8. The data represent the total consumption of energy for heating, cooling, lighting, auxiliary energy and equipment.

Table 8. Energy use and refrigerant indicators for defined outlet types.

Outlet Type	Heating (kWh/(m^2 year))	Electricity (kWh/(m^2 year))	Amount of Refrigerant (kg/m^2)	Annual Refrigerant Leakage (%)
Small store	114	463	0.3	10
Medium store	65	280	0.2	15
Large store	65	280	0.1	20
Wholesale market	37	172	-	-
Farmers' market	-	38	-	-

Source: own elaboration.

The energy consumption in medium and large stores has been taken as the average of the measured heat and electricity consumption in 203 outlets. The data for the small store were obtained based on a survey conducted in three stores. The energy consumption of wholesale markets and farmers' markets (mainly electricity for lighting) was taken from measured data from one similar outlet. Although energy demand data were obtained based on values from specific facilities, they represent European average values, and are consistent with the literature values [109,110]. Gas boilers and electricity grids were assumed as the energy sources for all considered buildings.

Transport Vehicles

Transportation is the second major activity in the food distribution chain and is therefore a key factor of environmental pressure. The LCA of transportation concerns, in particular, the manufacture and the use of vehicles to deliver and to carry food along the distribution chain.

As regards the manufacture phase, vehicles were divided into 4 types: truck (Heavy Goods Vehicle), delivery van (Light Goods Vehicle), passenger car and city bus. Table 9 summarizes the main assumptions for the calculation of the environmental impacts generated by the transportation phase.

Table 9. Assumptions applied to transportation vehicles.

Vehicle Type	Mass of the Vehicle (Tons)	Load Capacity (Tons)	Use of Fuel (l/100 km)	Maximum Mileage (km)
Heavy Goods Vehicle	8.2	21.00	33.0	1,000,000
Light Goods Vehicle	1.7	1.50	11.0	500,000
Passenger car	1.4	0.35 (5 persons)	8.0	250,000
City Bus	14.0	3.20 (45 persons)	32.0	1,300,000

Source: own elaboration.

The load capacities of vehicles were calculated as the average of a mix of means of transportation used by producers, based on data from farm surveys. Other parameters related to the flows of material and energy were assumed in relation to the characteristics of different types of vehicles.

With regard to passenger cars, environmental factors for the manufacturing and end of life phases were extrapolated from the environmental certificate provided by a reference car manufacturer [111]. Non-hazardous waste disposed was calculated separately, based on material quantities, using

LCA software. Emissions from delivery vans were estimated according to the emission level of a passenger car, taking into account the actual mass of the vehicle and the maximum mileage.

The environmental impacts of the manufacture phase for a city bus were assessed through the EPD documents concerning the same type of vehicle [112]. Emissions from the end of life phase were estimated using the environmental database provided by the LCA software. To estimate environmental indicators for heavy-duty vehicles, emission data for City Bus were recalculated to include the actual mass of the vehicle and the assumed average mileage in its lifetime.

Coefficients to calculate environmental indicators for all the transportation vehicles were finally converted to kilograms of emissions per travelled km. The applied environmental factors for analyzed transportation vehicles are shown in Table A1 in Appendix A.

Transportation—Use Phase Data

Transportation related eco-efficiency indicators were calculated considering P2R (farmers, intermediaries, couriers or retailers) and R2C (consumers using personal cars or public communication, which is represented in our analysis by city bus).

Concerning the vehicle use, the key data and information on P2R transportation was collected through the farm survey (means of transportation, use of fuel, locations and distances, average amount of goods transported).

Additional information on the typical organization of deliveries was used to calculate the Net Transportation Distance (NTD) attributed to specific distribution chains considered in the analysis. The NTD was calculated as the difference between a physical distance and the deductions made due to:

- the partial use of load capacity if other goods were transported;
- the proportion of the physical distance covered if other channels were served in one trip;
- the frequency of transporting goods (if any) on the return journey after delivering food products to the respective channel destination.

Specific transportation related parameters for the chain *"Sales to hypermarket chains"* were drawn from the study by Matulka [113].

Based on individual interviews from pilot studies conducted within the Strength2Food project and experts' knowledge, some additional assumptions were made regarding the transportation of products from wholesale markets to retail outlets.

Parameters for R2C transportation regarding the chains *a. Pick-your-own* and *b. On-farm sales to consumers* were estimated using primary information from producers. For the other types of distribution chains, information from interviews with consumers shopping on farmers' markets and experts' assumptions were used. The key data for LCA analyses relate the average weight of the consumers' basket in each type of the retail outlet and distances travelled with the use of a personal car or public transportation. For R2C transports, the Net Transportation Distance for consumers' travels was also assessed, in which physical distances from the retail outlets to consumers' place of living were accordingly adjusted (travels to buy food when travelling on other occasions than shopping (passing by)).

4. Results

4.1. Transportation Efforts Across Supply Chains

The basic transportation-related data are summarized in Table 10. The Net Food Miles indicator, expressed as the Net Transportation Distance in kilometers to transport 1 kilogram of purchased food, is considered one of the key factors influencing the environmental performance of chains.

Table 10. Food Miles and transportation-related data for food supply chains in the sample.

Supply Channels	Net Food Miles			Use of Fuel [L/kg]	
	Average [km/kg]	Of Which P2R [%]	Of Which R2C [%]	P2R	R2C
a. Pick-your-own	6.04	0.0	100.0	-	0.474
b. On-farm sales to consumers	3.75	0.0	100.0	-	0.300
c. Internet sales—courier deliveries	0.15	100.0	0.0	0.021	-
d. Direct deliveries to consumer	1.65	100.0	0.0	0.177	-
e. Sales on farmer's markets	1.36	32.1	67.9	0.040	0.058
f. Direct deliveries to retail	0.49	61.0	39.0	0.031	0.012
Short Chains Total	1.25	24.2	75.8	x	x
g. On-farm sales to intermediaries	0.26	29.2	70.8	0.016	0.012
h. Sales on wholesale market	0.62	68.6	31.4	0.056	0.012
i. Sales to hypermarket chains	0.67	24.7	75.3	0.021	0.036
Long Chains Total	0.53	41.0	63.4	x	x
Total	0.79	31.3	68.7	x	x

Source: own elaboration.

The results presented in Table 10 show that the average value of the Food Miles indicator for 'short' food supply chains (1.25 km/kg) is more than two times greater compared to 'long' chains (0.53), mainly due to the much smaller quantities transported in single deliveries. In long chains, the distances over which food is transported are much longer, but because of the large quantities transported, the indicator is significantly lower.

The vast majority of Food Miles are driven by consumers (68.7% in the sample on average). The highest Food Miles value characterizes chains with the full participation of consumers in transportation and small quantities transported (*a. Pick-your-own, b. On farm-sales*). The third largest, because of the small quantities transported, is the Food Miles indicator for *d. Direct deliveries to consumer*, followed by *e. Sales on farmer's markets*, mainly due to the location of markets at relatively long distances for both P2R and C2R travels.

Within long food supply chains, the proportional share of consumers involved in transportation differ greatly in *h. Sales on wholesale market* compared to other chains (*g. On-farm sales to intermediaries, i. Sales to hypermarket chains*), largely due to the use of mainly small vehicles for P2R transportation, both from the farm gate to a wholesale market and from wholesale to retail outlets. In the chains *g. On farm sales to intermediaries* and *i. Sales to hypermarket chains*, P2R transportation, mainly through the use of Heavy Goods vehicles, is much more effective; thus, consumers have a dominating share in Food Miles.

The use of fuel per kilogram of transported goods corresponds strongly with distances travelled, although depends also to a large extent on the type of car used, as well as on the quantities transported. The lowest fuel consumption was estimated for *c. Internet sales* (short chain) and within the category of long chains—*g. On-farm sales to intermediaries* and *i. Sales to hypermarket chains*. In all these channels, this is because of the large quantities transported in a single delivery, despite the relatively long distances travelled.

4.2. Environmental Impacts of Selected Supply Chains—LCA Results

The assessment of the environmental impacts of food supply chains according to the predetermined methodology of LCA includes three basic stages:

- impacts related to the production phase of devices, machines, vehicles and the construction of buildings used in the food distribution process,
- impacts related to the exploitation phase of specified devices and buildings (energy used in transportation, energy used for cooling, energy used to operate electrical devices etc.),
- impacts related to the waste management phase.

The selected eco-efficiency indicators that illustrate the environmental impacts of the analyzed food supply chains are presented in Table 11. All values express the amounts of emitted substances per 1 kg of product delivered through the chain.

Table 11. Eco-efficiency LCA indicators assessed for short and long food supply chains (kg emissions/kg of product).

Category of Supply Chains	Category of Environmental Impacts					
	Global Warming	Acidification	Eutrophication	Ozone Depletion Potential	Photochemical Oxidant Creation Potential	Non-Hazardous Waste Disposed
	kg CO_2e	10^{-3} kg SO_2e	10^{-4} kg PO_4e	10^{-7} kg CFC11e	10^{-4} kg C_2H_4e	10^{-2} kg
Short chains						
a. Pick-your-own	1.54	4.02	5.44	2.50	2.42	6.75
b. On-farm sales to consumers	0.97	2.54	3.44	1.58	1.53	4.27
c. Internet sales—courier deliveries	0.07	0.13	0.27	0.12	0.13	0.18
d. Direct deliveries to consumer	0.64	1.22	2.11	1.02	1.19	1.88
e. Sales on farmers' markets	0.49	1.58	2.25	0.69	1.03	4.44
f. Direct deliveries to retail	0.33	1.24	1.74	0.43	0.75	1.31
Long chains						
g. On-farm sales to intermediaries	0.24	0.92	1.41	0.30	0.54	1.56
h. Sales on wholesale market	0.40	1.29	2.00	0.55	0.85	1.80
i. Sales to hypermarket chains	0.33	1.15	1.74	0.46	0.70	2.29
AVERAGE	0.38	1.26	1.85	0.53	0.78	1.99

Source: own calculation.

In order to make the interpretation of results easier, the values of indicators are presented in a graphic form in Figure 3. The diagram illustrates the differences between the eco-efficiency indicators for the analyzed food supply chains presented in a relative way, as relations between specific parameters and the mean values of indicators for all chains in the sample. The mean was calculated as a weighted average, considering the volume of sales in respective chains.

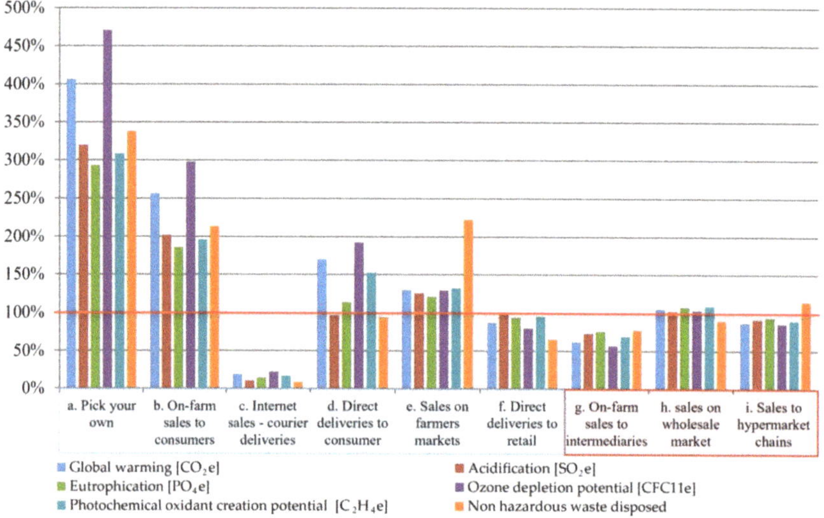

Figure 3. Relative eco-efficiency indicators (emissions per 1 kg of product) for supply chains related to weighted mean value = 100%. Source: own calculation.

The highest level of emissions for the eco-efficiency indicators characterizes *a. Pick-your-own* and *b. On-farm sales to consumers* (the only exception is the second highest "waste disposed" indicator), determined mainly by consumers' travels in both cases. Such a low environmental performance indicates the least effective use of individual cars as a means of transportation to transport relatively small quantities of food over long distances. Usually a trip to the farm in this form of shopping involves a purposeful journey, while the amount of produce transported is relatively small. Despite the very high values of all emissions, it is also worth paying attention to the proportionally high level of the "non-hazardous waste disposed" indicator, which results from a relatively large amount of waste generated by a passenger car in relation to the assumed total weight of products transported during its life time.

Emissions generated by *b. On-farm sales to consumers* are on average around 40% lower compared to *a. Pick-your-own*. This is due to the fact that, in the case of the first of these options, shopping is very often a spontaneous decision while passing by, which reduces the Net Transportation Distance. Traveling to *a. Pick-your-own* is usually a planned activity that involves a special trip to the farm. Even though quantities from pick-your-own are usually greater than from normal shopping, in the assessment of emissions, they do not counterbalance the longer distances travelled.

The chain *c. Internet sales—courier deliveries* is the least emissive, with values of all eco-efficiency indicators at the level of about 10% of the mean values for the sample. This is because the chain ensures the most efficient use of the means of transportation due to optimized distribution by courier companies. Even if individual deliveries are small, large quantities of all collected goods are transported with the use of heavy goods vehicles over considerable distances between collection points. Furthermore, the collection of parcels by couriers as well as last-mile deliveries to consumers are much more efficient in terms of environmental performance than the individual travels of consumers or producers. Another significant factor making Internet sales more eco-efficient is a lack of storage due to continuous, usually day-to-day, transportation.

The chain *d. Direct deliveries to consumer* is a "reversed" version of *b. On-farm sales to consumers*; however, it is environmentally much more efficient. In this case, transportation from the farm gate to consumers' place of living is done entirely by the producer, who transports greater loads in a single journey, usually serving a number of customers on one trip, often delivering food to collection points. Relatively high indicators of global warming potential and ozone depletion potential result from the fuel consumption, similar to chains *a.* and *b.*, but these are not sufficiently compensated by the greater loads transported.

Chains *e. Sales on farmers' markets* and *f. Direct deliveries to retail* differ from other short chains in that they require specific outlets for retail trade. The trade infrastructure on farmers' markets usually takes the form of simple construction stands spread over a paved and fenced area. Retail shops provide a limited area for highly concentrated sales and contain equipment, such as deep coolers or refrigerators, that are additional sources of emissions. A distinctive variable in the case of these two chains is the volume of sales from the trading area. In our study, it was 130 kg/m^2 of food products on farmers' markets and about 1100 kg/m^2 of food in retail shops on average.

As all estimated LCA indicators show, *e. Sales on farmers' markets* are noticeably less eco-efficient than *f. Direct deliveries to retail* shops. This is due to the rather extensively used trade infrastructure (lower sales from the trading area unit), more P2R transportation efforts (larger number of individual suppliers transporting relatively small quantities of products) and more consumer travels, because of the usually distant location of the majority of farmers' markets from consumers' place of living. In contrast to the latter, small- and medium-sized retail shops normally have much closer locations. Thus, shopping, often done in a passing-by situation on the way home or during travels for other purposes, involves much less transportation by car. This compensates for the usually lower weight of an average single purchase in a retail shop. Consumers shopping on farmers' markets usually buy more, but travel longer distances specifically for the purpose of doing so and more often with the use of a personal car.

All LCA indicators, except "waste disposed", which is very high in the case of *e. Sales on farmers' markets*, are about 20–25% higher than mean values for the sample and comparable eco-efficiency indicators for *f. Direct deliveries to retail*. The exception is the relatively high emission of harmful substances in this chain because of the more extensive use of infrastructure (significantly lower sales per square meter of sales area) and the shorter life of trading instalments.

The values of all eco-efficiency LCA indicators estimated for long supply chains are close to the mean values for the sample and comparable with *f. Direct deliveries to retail* for the short food supply chains. This is mainly due to the more effective use of trading facilities in some types of retail outlets (e.g., hypermarkets) and large volumes of goods transported in individual deliveries that make the means of transportation more effective (reducing energy consumption in the use phase).

Among the long chains, *h. Sales to wholesale market* are slightly less eco-efficient, which is due to the greater share of low-capacity cars used by farmers for transportation and thus higher values of Food Miles. Eco-efficiency indicators are further worsened because of transportation from wholesale to retail outlets. Although distances travelled on this leg are relatively short, small quantities and the frequency of transportation activities (very often every day in order to provide fresh produce) increase the value of eco-efficiency indicators.

Indicators for *i. Sales to hypermarket chains* are very similar to those characterizing the short chain *f. Direct deliveries to retail*. In the case of deliveries to hypermarkets, the distances travelled by products and also by consumers are longer, but the quantities transported are significantly greater, reducing the number of Food Miles, and vice versa; in *f. Direct deliveries to retail*, the distances travelled are much shorter, but smaller loads in a single delivery worsen the eco-efficiency indicators for this chain.

Figure 4 presents a synthetized comparison of the average values of eco-efficiency LCA indicators for the short and long supply chains analyzed in this study.

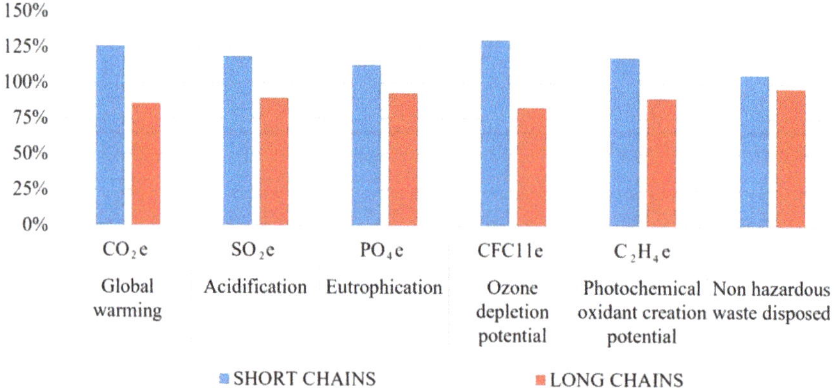

Figure 4. Average eco-efficiency indicators for short and long food supply chains related to weighted mean value (mean = 100%). Source: own calculation.

In Table 12, the relative share of transport in relation to the total values of the eco-efficiency indicators is presented. It can be observed that, as indicated by Browne et al. [77], in the entire emissions generated by food distribution through various chains, transport plays an important role.

Transportation is responsible for a huge 79% of the total carbon footprint in the case of 'short' food supply chains, and for about 53% in the case of 'long' supply chains. In all remaining indicators for SFSCs, the share of transport is quite high compared to long chains, ranging from a 22 percentage point difference in the case of ozone depletion, to a 33.4 percentage point difference in the case of acidification. This can be explained by the more direct way of selling food through short chains, where intermediaries and their facilities are not contributing to the emissions.

Table 12. Share of transport among eco-efficiency LCA indicators for short and long supply chains.

	Global Warming	Acidification	Eutrophication	Ozone Depletion Potential	Photochemical Oxidant Creation Potential	Non-Hazardous Waste Disposed
	kg CO_2e	10^{-3} kg SO_2e	10^{-4} kg PO_4e	10^{-7} kg CFC11e	10^{-4} kg C_2H_4e	10^{-2} kg
SFSCs	79.2%	66.3%	65.7%	85.3%	71.4%	54.8%
LFSCs	53.6%	32.9%	35.9%	63.3%	42.6%	30.8%
Sample	62.9%	45.0%	46.7%	71.3%	53.1%	39.5%

Source: own calculation.

In relation to the main hypothesis empirically tested in this paper, the results of our study reject the popular assumption that short food supply chains are more eco-efficient than long chains. It should be emphasized, however, that the values of eco-efficiency indicators display a large variability across analyzed chains, and especially across different types of SFSCs. A key reason for this is due to their specific features, such as the transport distance and the amount of goods per individual delivery. Moreover, the analysis shows that the environmental impacts of different food distribution systems are not only determined by the geographical distance between producer and consumer, but depend on numerous factors, including the supply chain infrastructure, the type and capacity of vehicles used, as well as specific conditions associated with food transportation, storage and the display of food in retail outlets (e.g., refrigerators and freezers, store shelves).

Our findings can hardly be compared in the context of the extant literature since, to the best of our knowledge, there are no similar, complex analyses made that quantify the eco-efficiency of several types of supply chains with the application of the LCA methodology. Opinions on the better environmental performance of short food supply chains are most often based on qualitative assessments. These assessments are typically lacking the required detailed information, or depth to the analysis, required to capture the complexity of food systems and, most importantly, are shaped by the simplifying assumption that shortening the travelling distances of products can significantly reduce the negative environmental externalities [11,15,18,19,37]. This view has, more recently, been frequently questioned [9–12,43–45].

The literature does contain examples of LCA-based eco-efficiency analysis of supply chains, although rather narrowly focused on single products and differently defined system boundaries, with a strong focus on production processes [69,114]. Such studies prove the applicability of the LCA methodology for assessing the environmental performance of food supply chains, providing a clear rationale for our work.

5. Conclusions

The food market today is rapidly evolving in terms of what is manifested, inter alia, by the development of new types of short supply chains. A widespread belief that SFSCs are more beneficial for the environment compared to conventional long chains is rejected by the results of our empirical testing with the use of the LCA approach. Shortening food supply chains in terms of organizational proximity does not automatically entail a more environmentally sustainable alternative to long supply chains nor mitigate some of the negative environmental externalities. Similarly, reducing the transportation distance from primary producers to end consumers, which is usually one of the specific features of SFSCs, may not lead to better eco-efficiency because of the small quantities typically transported in short supply chain deliveries.

More importantly, the results reveal that consumers make a significant contribution to the eco-efficiency measurement of all considered indicators. While this suggests a limited scope for reducing the environmental impacts of P2R transportation, there is significant potential for improving the eco-efficiency of supply chains through innovative business models for retail, which would not only shorten producer–consumer travel distances, but also make them more convenient and effective (e.g., last-mile delivery, group shopping on farmers' markets, internet sales, courier deliveries). Further research will be required, in this context, in order to identify and empirically assess the environmental benefits of existing/new business solutions, including organizational innovations in logistics.

While the LCA provides a suitable methodological tool for assessing the environmental impacts of different types of food chains, some important caveats must be noted. First, the analysis focuses on the distribution system, without examining the environmental impacts of the production stage. Second, potential selection bias implies that the sample cannot be considered representative of the whole population of farms across different countries. Finally, the modelling approach required several assumptions and simplifications to be made, with scientific implications for the universal validity of eco-efficiency indicators and the obtained results.

Our findings widen the path for further research in the area of supply chains. Empirical investigation may complement our study in terms of:

- Gaining deeper production data to validate our assumption on the use of technologies of production, irrespective of the choice of distribution channel;
- Searching for country-specific characteristics, such as the scale of operations, distribution organization, shopping patterns or energy mixes, which may differentiate the eco-efficiency of supply chains;
- Assuring the better representation of chains and the representativeness of research samples;
- Providing more detailed mapping and eco-efficiency assessments of food supply chains, including food processing.

The methodological and empirical contribution of this study represents, however, the first attempt in the literature that applies LCA analysis to a unique sample of coexistent short and long food supply chains, and thus contributes to the existing literature in terms of the observation of various food distribution paths and the assessment of their eco-efficiency. This study thus contributes to filling an existing gap in the empirical literature by providing a critical reflection on the realization of environmental benefits of SFSCs, which cannot be simply inferred via the association of close geographical proximity and short transport distances with low energy consumption, as typically suggested in qualitative studies.

The results of the study may contribute to the debate on the EU Green Deal. Our approach may be used for the creation of better suited, evidence-based environmental and energy policy instruments for supporting the most eco-efficient supply chains and innovative business models linking close primary producers and food consumers.

Author Contributions: Conceptualization, E.M., A.K., J.K., A.W.; methodology, E.M., A.K., J.K., A.M.-R., A.W.; software, A.K., J.K., A.W., M.G., K.P.; validation, E.M., A.K., J.K., A.W.; formal analysis, A.M.-R., and E.M.; data curation, E.M., A.K., J.K., A.M.-R., A.W., M.G., K.P., J.-L.L., B.T., Á.T., M.D., G.V.; writing—Original draft preparation, E.M., A.M.-R., A.K., J.K., P.S.; writing—Review and editing, E.M., A.K., J.K., A.M.-R., A.W., P.S., M.G., K.P., J.-L.L., B.T., Á.T., M.D., G.V.; project administration, A.M.-R. All authors have read and agreed to the published version of the manuscript.

Funding: This research received funding from the European Union's Horizon 2020 research and innovation program STRENGTH2FOOD under grant agreement no. 678024 and title: "Strengthening European Food Chain Sustainability by Quality and Procurement Policy".

Conflicts of Interest: The authors declare no conflict of interest. The funders had no role in the design of the study; in the collection, analyses, or interpretation of data; in the writing of the manuscript, or in the decision to publish the results.

Appendix A

Table A1. Environmental parameters for eco-efficiency assessments.

	Functional Unit	Global Warming Potential	Acidification Potential	Eutrophication Potential	Ozone Depletion Potential	Photo-Chemical Ozone Creation Potential	Non-Hazardous Waste Disposed
		GWP kg CO_{2e}	AP kg SO_{2e}	EP kg PO_{4e}	ODP kg CFC-11_e	POCP kg C_2H_{4e}	NHWD kg
Buildings and infrastructure							
Small store	unit/year	33,500	110	15	2.20×10^{-3}	6	900
Medium store	unit/year	246,100	650	98	1.35×10^{-2}	35	14,300
Large store	unit/year	1,028,000	3240	490	6.77×10^{-2}	173	71,600
Wholesales market	unit/year	3,196,600	15,460	2373	3.42×10^{-1}	884	452,200
Farmers' market	unit/year	111,800	590	85	1.08×10^{-2}	33	22,900
Transportation vehicles							
Passenger car	unit/km	0.0358	1.98×10^{-4}	1.17×10^{-5}	2.11×10^{-9}	1.67×10^{-5}	0.0098
VAN	unit/km	0.0432	2.39×10^{-4}	1.38×10^{-5}	2.53×10^{-9}	2.02×10^{-5}	0.0094
Truck	unit/km	0.0369	1.95×10^{-4}	1.07×10^{-4}	3.09×10^{-9}	2.45×10^{-5}	0.0089
City bus	unit/km	0.0449	2.48×10^{-4}	1.40×10^{-4}	3.97×10^{-9}	3.06×10^{-5}	0.0110
Energy sources							
Diesel	unit/l	3.240	4.67×10^{-3}	9.67×10^{-4}	5.50×10^{-7}	4.80×10^{-4}	0.019
Petrol 95E10	unit/l	2.800	6.00×10^{-3}	1.00×10^{-3}	5.00×10^{-7}	3.00×10^{-4}	0.020
Electricity, Europe	unit/kWh	0.386	2.17×10^{-3}	3.00×10^{-4}	4.17×10^{-8}	1.05×10^{-4}	0.014
Natural gas	unit/kWh	0.244	7.17×10^{-4}	4.83×10^{-5}	2.00×10^{-8}	4.83×10^{-5}	0.002

Source: own calculation.

References

1. World Resources Institute. *Climate Analysis Indicators Tool*; WRI: Washington, DC, USA, 2017.
2. Moriarty, P.; Wang, S.J. Eco-efficiency indicators for urban transport. *J. Sustain. Dev. Energy Water Environ. Syst.* **2015**, *3*, 183–195. [CrossRef]
3. European Commission. *Communication from the Commission to the European Parliament, the Council, the European Economic and Social Committee and the Committee of the Regions. A European Strategy for Low-Emission Mobility*; COM/2016/0501 Final; European Commission: Brussels, Belgium, 2016.
4. Zhang, S.; Witlox, F. Analyzing the impact of different transport governance strategies on climate change. *Sustainability* **2020**, *12*, 200. [CrossRef]
5. European Commission. Communication from the Commission to the European Parliament, the Council, the European Economic and Social Committee and the Committee of the Regions. A Policy Framework for Climate and Energy in the Period from 2020 to 2030. COM(2014) 15 Final; European Commission: Brussels, Belgium, 2014.
6. Pirog, R.S.; Van Pelt, T.; Enshayan, K.; Cook, E. *Food, Fuel, and Freeways: An Iowa Perspective on How Far Food Travels, Fuel Usage, and Greenhouse Gas Emissions*; Iowa State University: Ames, IA, USA, 2001. Available online: https://lib.dr.iastate.edu/cgi/viewcontent.cgi?article=1002&context=leopold_pubspapers (accessed on 13 September 2020).
7. European Commission. *Italian NRN Contribution to the "Short Supply Chain" Workshop*; EC: Bad Schandau, Germany, 2011; Available online: https://enrd.ec.europa.eu/enrd-static/fms/pdf/BAA4B613-92C1-DBB4-997B-7BFAFFA031CC.pdf (accessed on 13 September 2020).
8. Pretty, J. Some Benefits and Drawbacks of Local Food Systems. Bioregional Connections to Sustainable Foodsheds. Briefing Note for TVU/Sustain AgriFood Network. 2001. Available online: https://www.sustainweb.org/pdf/afn_m1_p2.pdf (accessed on 13 September 2020).
9. Shindelar, R. The Ecological Sustainability of Local Food Systems. In *Think Global, Eat Local: Exploring Foodways*; Pimbert, M., Shindelar, R., Schösler, H., Eds.; Rachel Carson Center: LMU Munich, Germany, 2015; pp. 19–23.
10. Galli, F.; Brunori, G. Short Food Supply Chains as Drivers of Sustainable Development Evidence Document; Laboratorio di studi rurali Sismondi, Italy. 2013. Available online: https://orgprints.org/28858/1/evidence-document-sfsc-cop.pdf (accessed on 13 September 2020).

11. Weber, C.L.; Matthews, H.S. Food-miles and the relative climate impacts of food choices in the United States. *Environ. Sci. Technol.* **2008**, *42*, 3508–3513. [CrossRef] [PubMed]
12. Malak-Rawlikowska, A.; Majewski, E.; Wąs, A.; Borgen, S.O.; Csillag, P.; Donati, M.; Freeman, R.; Hoàng, V.; Lecoeur, J.-L.; Mancini, M.C.; et al. Measuring the Economic, Environmental, and Social Sustainability of Short Food Supply Chains. *Sustainability* **2019**, *11*, 4004. [CrossRef]
13. Feenstra, G.W. Local food systems and sustainable communities. *Am. J. Altern. Agric.* **1997**, *12*, 28–36. [CrossRef]
14. Peters, R.; Markuszewska, A.; Prior, A.; Strano, A.; Bálint, B.; Midoux, B.; Bros, C.; Koutsaftaki, C.; Jochum, C.; Buffet, C.; et al. *Rural Review Issue 12—Local Food and Short Food Supply Chains, A Publication from the European Network of Rural Development-Online*; European Union: Brussels, Belgium, 2012. Available online: https://enrd.ec.europa.eu/sites/enrd/files/E8F24E08-0A45-F272-33FB-A6309E3AD601.pdf (accessed on 13 September 2020).
15. Kneafsey, M.; Venn, L.; Schmutz, U.; Balázs, B.; Trenchard, L.; Eyden-Wood, T.; Bos, E.; Foster, G.; Blackett, M. *Short Food Supply Chains and Local Food Systems in the EU. A State of Play of their Socio-Economic Characteristics*; Joint Research Centre (JRC): Luxembourg, 2013.
16. European Parliament. *Regulation (EU) No 1305/2013 of the European Parliament and of the Council of 17 December 2013 on Support for Rural Development by the European Agricultural Fund for Rural Development (EAFRD) and Repealing Council Regulation (EC)*; No 1698/2005; European Parliament: Brussels, Belgium, 2013.
17. Vittersø, G.; Torjusen, H.; Laitala, K.; Tocco, B.; Biasini, B.; Csillag, P.; de Labarre, M.D.; Lecoeur, J.-L.; Maj, A.; Majewski, E.; et al. Short Food Supply Chains and Their Contributions to Sustainability: Participants' Views and Perceptions from 12 European Cases. *Sustainability* **2019**, *11*, 4800. [CrossRef]
18. Sims, R.; Flammini, A.; Puri, M.; Bracco, S. *Opportunities for Agri-Food Chains to Become Energy-Smart*; FAO and USDA, U.S. Agency for International Development: Washington, DC, USA, 2015. Available online: http://www.fao.org/3/a-i5125e.pdf (accessed on 13 September 2020).
19. Bomford, M. Beyond Food Miles—Resilience. Available online: https://www.resilience.org/stories/2011-03-09/beyond-food-miles/ (accessed on 6 July 2020).
20. Monforti-Ferrario, F.; Dallemand, J.-F.; Pascua, I.P.; Motola, V.; Banja, M.; Scarlat, N.; Medarac, H.; Castellazzi, L.; Labanca, N.; Bertoldi, P.; et al. *Energy Use in the EU Food Sector: State of Play and Opportunities for Improvement*; Joint Research Centre (JRC): Luxembourg, 2015. Available online: https://publications.jrc.ec.europa.eu/repository/bitstream/JRC96121/ldna27247enn.pdf (accessed on 6 July 2020).
21. Ladha-Sabur, A.; Bakalis, S.; Fryer, P.J.; Lopez-Quiroga, E. Mapping energy consumption in food manufacturing. *Trends Food Sci. Technol.* **2019**, *86*, 270–280. [CrossRef]
22. Fisher, D.; McKinnon, A.; Palmer, A. Reducing the external costs of food distribution in the UK. In *Delivering Performance in Food Supply Chains*; Mena, C., Stevens, G., Eds.; Woodhead Publishing Limited: Oxford, UK, 2010; pp. 459–477, ISBN 9781845694715.
23. Foucherot, C.; Rogissart, L. Estimating Greenhouse Gas Emissions from Food Consumption: Methods and Results. Available online: https://www.i4ce.org/download/estimating-greenhouse-gas-emissions-from-food-consumption-methods-and-results/ (accessed on 13 August 2020).
24. FAO. *Energy-Smart Food for People and Climate*; Issue Paper; Food and Agriculture Organization of the United Nations: Rome, Italy, 2011.
25. Akkerman, R.; Farahani, P.; Grunow, M. Quality, safety and sustainability in food distribution: A review of quantitative operations management approaches and challenges. *OR Spectr.* **2010**, *32*, 863–904. [CrossRef]
26. European Commission. *Guidance on the Implementation of Articles 11, 12, 14, 17, 18, 19 and 20 of Regulation (EC) N° 178/2002 on General Food Law. Conclusions of the Standing Committee on the Food Chain and Animal Health*; EC: Brussels, Belgium, 2010.
27. Pal, A.; Kant, K. A Food Transportation Framework for an Efficient and Worker-Friendly Fresh Food Physical Internet. *Logistics* **2017**, *1*, 10. [CrossRef]
28. Peck, H. *Resilience in the Food Chain: A Study of Business Continuity Management in the Food and Drink Industry. Final Report to the Department for Environment, Food and Rural Affairs*; DEFRA: London, UK, 2006.
29. Commitee on World Food Security. *Global Strategic Framework for Food Security and Nutrition (GSF)*; FAO: Rome, Italy, 2017. Available online: http://www.fao.org/3/MR173EN/mr173en.pdf (accessed on 13 August 2020).

30. Ziska, L.; Crimmins, A.; Auclair, A.; DeGrasse, S.; Garofalo, J.F.; Khan, A.S.; Loladze, I.; de León, A.A.P.; Showler, A.; Thurston, J.; et al. Food Safety, Nutrition, and Distribution. In *The Impacts of Climate Change on Human Health in the United States: A Scientific Assessment*; U.S Global Change Research Program: Washington, DC, USA, 2016; pp. 189–216. ISBN 9780309377270. Available online: http://medsocietiesforclimatehealth.org/wp-content/uploads/2016/11/Health-Impacts-of-Climate-Change-in-the-U.S.-2016_Full-Report_small.pdf (accessed on 13 August 2020).
31. Candel, J.J.L. Diagnosing integrated food security strategies. *NJAS Wagening. J. Life Sci.* **2018**, *84*, 103–113. [CrossRef]
32. Chiffoleau, Y.; Millet-Amrani, S.; Canard, A. From Short Food Supply Chains to Sustainable Agriculture in Urban Food Systems: Food Democracy as a Vector of Transition. *Agriculture* **2016**, *6*, 57. [CrossRef]
33. Malak-Rawlikowska, A.; Majewski, E.; Wąs, A.; Gołaś, M.; Kłoczko-Gajewska, A.; Borge, S.O.; Coppola, E.; Csillag, P.; de Labarre, M.D.; Freeman, R.; et al. *Quantitative Assessment of Economic, Social and Environmental Sustainability of Short Food Supply Chains and Impact on Rural Territories*; Deliverable 7.2., Strength2Food Project no.678024; Strength2Food: Krakow, Poland, 2019. Available online: https://www.strength2food.eu/wp-content/uploads/2019/02/D7.2-Quantitative-assessment-of-economic-social-and-environmental-sustainability-of-short-food-supply-chains-and-impact-on-rural-territories_final_protected.pdf (accessed on 13 August 2020).
34. Marsden, T.; Banks, J.; Bristow, G. Food supply chain approaches: Exploring their role in rural development. *Sociol. Rural.* **2000**, *40*, 424–438. [CrossRef]
35. Renting, H.; Marsden, T.K.; Banks, J. Understanding alternative food networks: Exploring the role of short food supply chains in rural development. *Environ. Plan. A* **2003**, *35*, 393–411. [CrossRef]
36. Fondse, M. Grown Close to Home: A Typology of Short Food Supply Chain Business Models in the Netherlands. Master's Thesis, Wageningen University, Wageningen, The Netherlands, 2012.
37. Erker, T.; Louis, S. Sustainability of Local Food Systems: Practices and Motivations of Farmers at St. Louis Farmers Markets. Senior Honors Thesis, Washington University in St. Louis, St. Louis, MI, USA, 2010.
38. Aubert, M. The Determinants of Selling through a Short Food Supply Chains: An Application to the French Case; Post-Print hal-01296422, HAL. 2015. Available online: https://ideas.repec.org/p/hal/journl/hal-01296422.html (accessed on 13 August 2020).
39. Conner, D.; Colasanti, K.; Ross, R.B.; Smalley, S.B. Locally grown foods and farmers markets: Consumer attitudes and behaviors. *Sustainability* **2010**, *2*, 742–756. [CrossRef]
40. Wittman, H.; Beckie, M.; Hergesheimer, C. Linking Local Food Systems and the Social Economy? Future Roles for Farmers' Markets in Alberta and British Columbia. *Rural Sociol.* **2012**, *77*, 36–61. [CrossRef]
41. Galli, F.; Bartolini, F.; Brunori, G.; Colombo, L.; Gava, O.; Grando, S.; Marescotti, A. Sustainability assessment of food supply chains: An application to local and global bread in Italy. *Agric. Food Econ.* **2015**, *3*, 21. [CrossRef]
42. Mundler, P.; Laughrea, S. The contributions of short food supply chains to territorial development: A study of three Quebec territories. *J. Rural Stud.* **2016**, *45*, 218–229. [CrossRef]
43. Gonçalves, A.; Zeroual, T. Logistic Issues and Impacts of Short Food Supply Chains: Case Studies in Nord—Pas de Calais, France. In *Toward Sustainable Relations between Agriculture and the City*; Soulard, C.-T., Perrin, C., Valette, E., Eds.; Springer: Berlin/Heidelberg, Germany, 2017; pp. 33–49.
44. Mancini, M.; Menozzi, D.; Donati, M.; Biasini, B.; Veneziani, M.; Arfini, F. Producers' and Consumers' Perception of the Sustainability of Short Food Supply Chains: The Case of Parmigiano Reggiano PDO. *Sustainability* **2019**, *11*, 721. [CrossRef]
45. Bloemhof, J.M.; Soysal, M. Sustainable Food Supply Chain Design. In *Sustainable Supply Chains*; Bouchery, Y., Corbett, C.J., Fransoo, J.C., Tan, T., Eds.; Springer: Amsterdam, The Netherlands, 2017; pp. 395–412.
46. Tassou, S.A.; De-Lille, G.; Ge, Y.T. Food transport refrigeration—Approaches to reduce energy consumption and environmental impacts of road transport. *Appl. Therm. Eng.* **2009**, *29*, 1467–1477. [CrossRef]
47. Sims, R.; Schaeffer, R.; Creutzig, F.; Cruz-Núñez, X.; D'Agosto, M.; Dimitriu, D.; Figueroa-Meza, M.J.; Fulton, L.; Kobayashi, S.; Lah, O.; et al. Transport. In *Climate Change 2014: Mitigation of Climate Change*; Edenhofer, O., Pichs-Madruga, R., Sokona, Y., Farahani, E., Kadner, S., Seyboth, K., Adler, A., Baum, I., Brunner, S., Eickemeier, P., et al., Eds.; Cambridge University Press: Cambridge, UK, 2014; pp. 599–670.
48. Nielsen, J.Ø.; Vigh, H. Adaptive lives. Navigating the global food crisis in a changing climate. *Glob. Environ. Chang.* **2012**, *22*, 659–669. [CrossRef]

49. Tirado, M.C.; Clarke, R.; Jaykus, L.A.; McQuatters-Gollop, A.; Frank, J.M. Climate change and food safety: A review. *Food Res. Int.* **2010**, *43*, 1745–1765. [CrossRef]
50. OECD. *Enhancing the Mitigation of Climate Change though Agriculture*; OECD: Paris, France, 2019.
51. Gowreesunker, B.L.; Tassou, S.A. Approaches for modelling the energy flow in food chains. *Energy. Sustain. Soc.* **2015**, *5*, 7. [CrossRef]
52. WBCSD; Lehni, M. *Eco-Efficiency. Creating More Value with Less Impact*; WBCSD: Geneva, Switzerland, 2000.
53. Gómez-Limón, J.A.; Picazo-Tadeo, A.J.; Reig-Martínez, E. Eco-efficiency assessment of olive farms in Andalusia. *Land Use Policy* **2012**, *29*, 395–406. [CrossRef]
54. Schaltegger, S.; Burritt, R. *Contemporary Environmental Accounting. Issues, Concepts and Practice*; Routledge: London, UK, 2000.
55. Rybaczewska-Błażejowska, M.; Gierulski, W. Eco-Efficiency Evaluation of Agricultural Production in the EU-28. *Sustainability* **2018**, *10*, 4544. [CrossRef]
56. Huppes, G.; Ishikawa, M. A framework for quantified eco-efficiency analysis. *J. Ind. Ecol.* **2005**, *9*, 25–41. [CrossRef]
57. Müller, K.; Holmes, A.; Deurer, M.; Clothier, B.E. Eco-efficiency as a sustainability measure for kiwifruit production in New Zealand. *J. Clean. Prod.* **2015**, *106*, 333–342. [CrossRef]
58. Caiado, R.G.G.; Dias, R.; Mattos, L.V.; Quelhas, O.L.G.; Filho, W.; Dias, R.d.F.; Mattos, L.V.; Quelhas, O.L.G.; Filho, W.L. Towards sustainable development through the perspective of eco-efficiency—A systematic literature review. *J. Clean. Prod.* **2017**, *165*, 890–904. [CrossRef]
59. World Business Council for Sustainable Development. *Eco-Efficient Leadership for Improved Economic and Environmental Performance*; WBCSD: Geneva, Switzerland, 1996.
60. Maxime, D.; Marcotte, M.; Arcand, Y. Development of eco-efficiency indicators for the Canadian food and beverage industry. *J. Clean. Prod.* **2006**, *14*, 636–648. [CrossRef]
61. OECD. *Ecoefficiency*; OECD Publishing: Paris, France, 1998.
62. UN. Eco-Efficiency Indicators: Measuring Resource-Use Efficiency and the Impact of Economic Activities on the Environment. Available online: https://sustainabledevelopment.un.org/index.php?page=view&type=400&nr=785&menu=1515 (accessed on 18 August 2020).
63. Burchart-Korol, D.; Kruczek, M.; Czaplicka-Kolarz, K. Wykorzystanie ekoefektywności w ocenie poziomu ekoinnowacyjności. In *Innowacje w Zarządzaniu i Inżynierii Produkcji*; Oficyna Wydawnicza Polskiego Towarzystwa Zarządzania Produkcją, Poland: Opole, Poland, 2013.
64. Saling, P.; Kicherer, A.; Dittrich-Krämer, B.; Wittlinger, R.; Zombik, W.; Schmidt, I.; Schrott, W.; Schmidt, S. Eco-efficiency analysis by BASF: The method. *Int. J. Life Cycle Assess.* **2002**, *7*, 203–218. [CrossRef]
65. Pishgar-Komleh, S.H.; Zylowski, T.; Rozakis, S.; Kozyra, J. Efficiency under different methods for incorporating undesirable outputs in an LCA+DEA framework: A case study of winter wheat production in Poland. *J. Environ. Manag.* **2020**, *260*, 110–138. [CrossRef]
66. Charles, R.; Jolliet, O.; Gaillard, G.; Pellet, D. Environmental analysis of intensity level in wheat crop production using life cycle assessment. *Agric. Ecosyst. Environ.* **2006**, *113*, 216–225. [CrossRef]
67. Wang, M.; Wu, W.; Liu, W.; Bao, Y. Life cycle assessment of the winter wheat-summer maize production system on the North China Plain. *Int. J. Sustain. Dev. World Ecol.* **2007**, *14*, 400–407. [CrossRef]
68. Liang, L.; Chen, Y.; Gao, W.; Sui, P.; Chen, D.; Zhang, W. Life cycle environmental impact assessment in winter wheat-summer maize system in north China plain. *J. Agric. Environ. Sci.* **2009**, *28*, 1773–1776.
69. McCarthy, D.; Matopoulos, A.; Davies, P. Life cycle assessment in the food supply chain: A case study. *Int. J. Logist. Res. Appl.* **2015**, *18*, 140–154. [CrossRef]
70. Mesaric, J.; Franjkovic, J.; Šebalj, D. Supply chains in the context of Life Cycle Assessment and Sustainability. In Proceedings of the 16th International Scientific Conference Business Logistics in Modern Management, Osijek, Croatia, 13 October 2016.
71. Carlsson-Kanyama, A. Climate change and dietary choices—How can emissions of greenhouse gases from food consumption be reduced? *Food Policy* **1998**, *23*, 277–293. [CrossRef]
72. Anand, C.K.; Amor, B. Recent developments, future challenges and new research directions in LCA of buildings: A critical review. *Renew. Sustain. Energy Rev.* **2017**, *67*, 408–416. [CrossRef]
73. van der Giesen, C.; Cucurachi, S.; Guinée, J.; Kramer, G.J.; Tukker, A. A critical view on the current application of LCA for new technologies and recommendations for improved practice. *J. Clean. Prod.* **2020**, *259*, 120904. [CrossRef]

74. Eriksson, E.; Blingeb, M.; Liivgren, G. Life cycle assessment of the road transport sector. *Sci. Total Environ.* **1996**, *189/190*, 69–76. [CrossRef]
75. Li, C.; Cui, S.P.; Gong, X.Z.; Meng, X.C.; Sun, B.X.; Liu, Y. Life Cycle Assessment of Heavy-Duty Truck for Highway Transport in China. *J. Chem. Inf. Model.* **2013**, *53*, 1689–1699. [CrossRef]
76. Yang, L.; Hao, C.; Chai, Y. Life Cycle Assessment of Commercial Delivery Trucks: Diesel, Plug-In Electric, and Battery-Swap Electric. *Sustainability* **2018**, *10*, 4547. [CrossRef]
77. Browne, M.; Rizet, C.; Anderson, S.; Allen, J.; Keïta, B. Life cycle assessment in the supply chain: A review and case study. *Transp. Rev.* **2005**, *25*, 761–782. [CrossRef]
78. EEA. *Electric Vehicles from Life Cycle and Circular Economy Perspectives*; TERM 2018: Transport and Environment Reporting Mechanism (TERM) Report; European Environment Agency, Publications Office of the European Union: Luxembourg, 2018. Available online: file:///C:/Users/Agata%20i%20Marcin/Downloads/TH-AL-18-012-EN-N%20TERM%202018.pdf (accessed on 13 August 2020).
79. Egede, P.; Dettmer, T.; Herrmann, C.; Kara, S. Life cycle assessment of electric vehicles—A framework to consider influencing factors. *Procedia CIRP* **2015**, *29*, 233–238. [CrossRef]
80. Patterson, J.; Johnson, A. Understanding the Life Cycle GHG Emissions for Different Vehicle Types and Powertrain Technologies. Final Report for LowCVP; Ricardo plc. 2018. Available online: file:///C:/Users/Agata%20i%20Marcin/Downloads/LowCVP-LCA_Study-Final_Report.pdf (accessed on 13 August 2020).
81. Syed, A.; Van Mierlo, J.; Messagie, M. Life cycle assessment of electrification of heavy-duty vehicle. In Proceedings of the 32nd International Electric Vehicle Symposium, Lyon, France, 19–22 May 2019; pp. 1–12.
82. Nordelöf, A.; Romare, M.; Tivander, J. Life cycle assessment of city buses powered by electricity, hydrogenated vegetable oil or diesel. *Transp. Res. Part D Transp. Environ.* **2019**, *75*, 211–222. [CrossRef]
83. Cabeza, L.F.; Rincón, L.; Vilariño, V.; Pérez, G.; Castell, A. Life cycle assessment (LCA) and life cycle energy analysis (LCEA) of buildings and the building sector: A review. *Renew. Sustain. Energy Rev.* **2014**, *29*, 394–416. [CrossRef]
84. Nwodo, M.N.; Anumba, C.J. A review of life cycle assessment of buildings using a systematic approach. *Build. Environ.* **2019**, *162*. [CrossRef]
85. Bahramian, M.; Yetilmezsoy, K. Life cycle assessment of the building industry: An overview of two decades of research (1995–2018). *Energy Build.* **2020**, *219*, 109917. [CrossRef]
86. Vilches, A.; Garcia-Martinez, A.; Sanchez-Montañes, B. Life cycle assessment (LCA) of building refurbishment: A literature review. *Energy Build.* **2017**, *135*, 286–301. [CrossRef]
87. Lavagna, M.; Baldassarri, C.; Campioli, A.; Giorgi, S.; Valle, A.d.; Castellani, V.; Sala, S. Benchmarks for environmental impact of housing in Europe: Definition of archetypes and LCA of the residential building stock. *Build. Environ.* **2018**, *145*, 260–275. [CrossRef]
88. Schlegl, F.; Gantner, J.; Traunspurger, R.; Albrecht, S.; Leistner, P. LCA of buildings in Germany: Proposal for a future benchmark based on existing databases. *Energy Build.* **2019**, *194*, 342–350. [CrossRef]
89. Komerska, A.; Kwiatkowski, J. Preliminary Study on the GWP Benchmark of Office Buildings in Poland Using the LCA Approach. *Energies* **2020**, *13*, 3298.
90. Chen, Y.; Ng, S.T. Factoring in embodied GHG emissions when assessing the environmental performance of building. *Sustain. Cities Soc.* **2016**, *27*, 244–252. [CrossRef]
91. Zabalza, I.; Scarpellini, S.; Aranda, A.; Llera, E.; Jáñez, A. Use of LCA as a tool for building ecodesign. A case study of a low energy building in Spain. *Energies* **2013**, *6*, 3901–3921. [CrossRef]
92. Lausselet, C.; Borgnes, V.; Brattebø, H. LCA modelling for Zero Emission Neighbourhoods in early stage planning. *Build. Environ.* **2019**, *149*, 379–389. [CrossRef]
93. Basbagill, J.; Flager, F.; Lepech, M.; Fischer, M. Application of life-cycle assessment to early stage building design for reduced embodied environmental impacts. *Build. Environ.* **2013**, *60*, 81–92. [CrossRef]
94. Ramesh, T.; Prakash, R.; Shukla, K.K. Life cycle energy analysis of buildings: An overview. *Energy Build.* **2010**, *42*, 1592–1600. [CrossRef]
95. Saade, M.R.M.; Guest, G.; Amor, B. Comparative whole building LCAs: How far are our expectations from the documented evidence? *Build. Environ.* **2020**, *167*, 106449. [CrossRef]
96. Kalluri, S. Comparative Life Cycle Assessment of Road and Multimodal Transportation Options—A Case Study of Copperwood Project. Master's Thesis, Michigan Technological University, Houghton, MI, USA, 2016.

97. Weber, C.L.; Hendrickson, C.T.; Matthews, H.S.; Nagengast, A.; Nealer, R.; Jaramillo, P. Life cycle comparison of traditional retail and e-commerce logistics for electronic products: A case study of buy.com. In Proceedings of the 2009 IEEE International Symposium on Sustainable Systems and Technology (ISSST '09) in Cooperation with 2009 IEEE International Symposium on Technology and Society (ISTAS), Tempe, AZ, USA, 18–20 May 2009.
98. Michelson, H.; Boucher, S.; Cheng, X.; Huang, J.; Jia, X. Connecting supermarkets and farms: The role of intermediaries in Walmart China's fresh produce supply chains. *Renew. Agric. Food Syst.* **2017**, *33*, 47–59. [CrossRef]
99. Ferreira, K.; Goh, J.; Valavi, E. *Intermediation in the Supply of Agricultural Products in Developing Economies;* Harvard Business School Research Paper Series; Harvard Business School: Boston, MA, USA, 2017. [CrossRef]
100. Cadilhon, J.; Fearne, A.; Hughes, D.; Moustier, P. *Wholesale Markets and Food Distribution in Europe: New Strategies for Old Functions;* Discussion Paper; Centre for Food Chain Research, Imperial College: London, UK, 2003.
101. International Organization for Standardization. *ISO 14040: Environmental Management—Life Cycle Assessment—Principles and Framework;* IOS: Geneva, Switzerland, 2006. Available online: https://www.iso.org/standard/37456.html (accessed on 13 August 2020).
102. International Organization for Standardization. *ISO 14044. Environmental Management—Life Cycle Assessement—Requirements and Guidelines;* IOS: Geneva, Switzerland, 2006; Available online: https://www.iso.org/standard/38498.html (accessed on 13 August 2020). [CrossRef]
103. Bionova LTD. One Click LCA. Available online: https://www.oneclicklca.com/ (accessed on 13 August 2020).
104. European Committee for Standardization (CEN). *EN 15978. Sustainability of Construction Works—Assessment of Environmental Performance of Buildings—Calculation Method;* European Committee for Standardization: Brussels, Belgium, 2011.
105. International Organization for Standardization. *ISO 14025. Environmental Labels and Declarations—Type III Environmental Declarations—Principles and Procedures;* IOS: Geneva, Switzerland, 2006.
106. Central Statistical Office. *Rynek wewnętrzny w 2018 roku;* Główny Urząd Statystyczny: Warsaw Poland, 2019.
107. Vukovic, Z.; Regodic, D. Freon Leakage. *Int. J. Sci. Res.* **2014**, *4*, 1–9.
108. Koronaki, I.P.; Cowan, D.; Maidment, G.; Beerman, K.; Schreurs, M.; Kaar, K.; Chaer, I.; Gontarz, G.; Christodoulaki, R.I.; Cazauran, X. Refrigerant emissions and leakage prevention across Europe—Results from the RealSkillsEurope project. *Energy* **2012**, *45*, 71–80. [CrossRef]
109. Kolokotroni, M.; Mylona, Z.; Evans, J.; Foster, A.; Liddiard, R. Supermarket energy use in the UK. *Energy Procedia* **2019**, *161*, 325–332. [CrossRef]
110. Tassou, S.A.; Ge, Y.; Hadawey, A.; Marriott, D. Energy consumption and conservation in food retailing. *Appl. Therm. Eng.* **2011**, *31*, 147–156. [CrossRef]
111. Mercedes-Benz. *Environmental Cycle, Certificate Mercedes-Benz E-Class;* Daimler AG, Untertürkheim Department: Group Environmental Protection, RD/RSE; Mercedes-Benz: Stuttgart, Germany, 2015.
112. EPD International AB. *Environmental Product Declaration, Irizar I4 Coach;* S-P-01571; IRIZAR S.COOP: Gipuzkoa, Spain, 2019.
113. Matulka, P. Carbon FootPrint dla wybranych łańcuchów dostaw artykułów żywnościowych do sieci hipermarketów. Master's Thesis, Warsaw University of Life Sciences, Warsaw, Poland, 2018.
114. Andersson, K.; Ohlsson, T. Life cycle assessment of bread produced on different scales. *Int. J. Life Cycle Assess.* **1999**, *4*, 25–40. [CrossRef]

© 2020 by the authors. Licensee MDPI, Basel, Switzerland. This article is an open access article distributed under the terms and conditions of the Creative Commons Attribution (CC BY) license (http://creativecommons.org/licenses/by/4.0/).

Review

A Bibliometric Analysis of Carbon Labeling Schemes in the Period 2007–2019

Rui Zhao [1,*], Dingye Wu [1] and Sebastiano Patti [2]

1 Faculty of Geosciences and Environmental Engineering, Southwest Jiaotong University, Chengdu 611756, China; wuu0715@163.com
2 Department of Business and Economics, University of Catania, 95131 Catania, Italy; spatti@unict.it
* Correspondence: ruizhaoswjtu@hotmail.com; Tel.: +86-6636-7628

Received: 23 June 2020; Accepted: 14 August 2020; Published: 16 August 2020

Abstract: Carbon labeling schemes enable consumers to be aware of carbon emissions regarding products or services, to help change their purchasing behaviors. This study provides a bibliometric analysis to review the research progress of carbon labeling schemes during the period 2007–2019, in order to provide insight into its future development. Number of publications, countries of publications, authors, institutions, and highly cited papers are included for statistical analysis. The CiteSpace software package is used to visualize the national collaboration, keywords co-appearance, and aggregation. The results are given as follows: (1) there are 175 articles published in the pre-defined period, which shows a gradual increase, with a peak occurred in 2016; (2) carbon labeling schemes are mainly applied to grocery products, and gradually emerged in construction and tourism. (3) Existing studies mainly focus on examination of utility of carbon labeling schemes, by conducting surveys to investigate individual perception, preference, and willingness to pay. (4) Future research will include the optimization of life cycle assessment for labeling accreditation, improvement of labeling visualization for better expression, and normalization of various environmental labels to promote sustainable consumption.

Keywords: carbon label; bibliometric analysis; CiteSpace; carbon labeling scheme; purchase intention; willingness to pay

1. Introduction

Greenhouse gas (GHG) emissions have given rise to global warming, which has aroused world concerns to adopt mitigation strategies to promote sustainable development [1]. Since the Paris Climate Summit in 2015, more than 160 countries have formulated policies regarding sustainable consumption and production to achieve emission reduction targets [2]. Among them, carbon labeling scheme is an insightful policy tool, by revealing lifecycle-based carbon emissions of a product or service, to encourage transition of low carbon consumption and production [3]. The first carbon labeling scheme was designed in 2006 by UK Carbon Trust [1]. One of its essential measures is to alleviate the impact of carbon emissions on the UK's sectors of production and services [4]. More than 2000 products with over 90 international brands have been issued by Carbon Trust to implement the scheme [5]. Since then, a number of developed countries (such as the U.S., France, Switzerland, Japan, and Canada) adopted such a policy to reveal information of environmental impact regarding a product or service [6]. Such scheme is intended to help consumers be conscious of climate change, to change their purchasing behaviors [7]. Consumer pro-environmental purchasing behavior is an important factor to promote decarbonization [8]. Over 80% of online surveyed consumers from eight countries admitted that they support carbon labeling policy as a useful tool to mitigate climate change [9]. Recent surveys further identified that consumers prefer purchasing carbon labeled products, and they are willing to pay a premium of up to 20% of the original price [10]. However, purchasing behavior is a complex decision-making process, which may differ significantly from gender, age, education, and income level, to result in unintended consequences on market demand of carbon labeled products [7,11,12]. Whether the carbon labeling scheme triggers consumers to take pro-environmental purchasing actions lays room for future studies.

In addition, carbon labeling is a voluntary behavior for enterprises or organizations [13]. One of the main motivations about carbon labeling regards marketing strategies, but carbon labeling may enhance consumption behavior towards environmental sustainability. Although such a scheme provides an opportunity to take a holistic carbon accounting on product supply chain and improve its green performance, there is a lack of sufficient incentives to drive its implementation [14]. Since most of the enterprises are driven by profitable motivation, they may be unwilling to attempt carbon labeling due to cost and benefit consideration, including the certification cost, market risk, governmental policy orientation, etc. [15,16]. For instance, China currently has not put a carbon labeling policy into wide practice, which exerts limited influences on market prospect [11,17].

In such context, it is implied that the carbon labeling scheme is still in dispute. A comprehensive review regarding its progress is essential, through which the research hotspots can be identified to lay out foundation for future study. Common literature review is based upon deliberate screening, which may not fully address the advances in a specific research area [18]. This study provides a holistic review on the research progress of carbon labeling schemes in the period 2007–2019 by using a bibliometric analysis. A number of selected indicators, e.g., number of publications, countries, authors, institutions, highly cited papers, and keywords have been incorporated into the analysis to highlight the research hotspots. Moreover, CiteSpace software package is employed to visualize the structures and connections in the retrieved scientific literatures [19]. The scope of this work is to define a comprehensive and holistic review on the development of carbon labeling schemes, through which research hotspots can be identified to lay out a foundation for future study. It is expected that this study may provide insight into the future development of carbon labeling schemes. The specific contributions of this study are: providing an alternative approach to review the research progress of carbon labeling based on bibliometrics, identifying the research trends through an analysis of descriptive statistics on the bibliometric indicators, laying out a foundation for future development of carbon labeling schemes.

The rest of the paper is given as follows: Section 2 introduces the data sources and method for data cleaning, Section 3 presents the results regarding bibliometric indicators, including number of publications, countries, subject categories, published journals, distribution of institution, highly cited papers, and keywords. Section 4 summarizes the paper, discusses the challenges towards carbon labeling schemes, and presents implications.

2. Data and Methods

The literature data were obtained by a predefined information retrieval from the Web of Science (WOS) Core Collection database in the period of 2007–2019, specifically from the sub-databases of Science Citation Index Expanded, Social Sciences Citation Index, and Arts and Humanities Citation Index. The predefined entries were wet based upon the "carbon label*" and the highlighted words in its associated definition, e.g., environmental impact, life cycle assessment, product, service, etc. There were 2016 records at the retrieval time 14/5/2020, and their document types were confined to Article or Review in English, as shown in Table 1. However, most of the publications focus on analytical chemistry, physical chemistry, biochemistry molecular biology, microbiology, etc., which are apparently not conform to the research topic. After removal of the duplicated and irrelevant publications, there were 175 articles finally identified for further bibliometric analysis.

The retrieval results were performed by descriptive statistics corresponding to a set of bibliometric indicators, such as number of publications, country, category, journal, institution, authors, highly cited papers, and keywords, to investigate the attentions to the carbon label related studies. Specifically, Microsoft Excel and bibliometric online analysis platforms were applied to implement the statistical analysis. The values corresponding to journal impact factors were from Journal Citation Reports in 2018. CiteSpace software package was used to visualize the underlying connections among keywords by generation of a co-occurrence network.

Table 1. Criteria for literature data retrieval.

Set	Results	Search Criteria
#6	2016	(#5 OR #4 OR #3 OR #2 OR #1) *AND* **LANGUAGE:** (English) *AND* **DOCUMENT TYPES:** (Article OR Review) *Indexes=SCI-EXPANDED, SSCI, A&HCI TIMESPAN=2007-2019*
#5	210	(TS=(carbon label*) AND TS=(environmental impact)) *AND* **LANGUAGE:** (English) *AND* **DOCUMENT TYPES:** (Article OR Review) *Indexes=SCI-EXPANDED, SSCI, A&HCI TIMESPAN=2007-2019*
#4	378	(TS=(carbon label*) AND TS=(consume*)) *AND* **LANGUAGE:** (English) *AND* **DOCUMENT TYPES:** (Article OR Review) *Indexes=SCI-EXPANDED, SSCI, A&HCI TIMESPAN=2007-2019*
#3	68	(TS=(carbon label*) AND TS=(life cycle assessment)) *AND* **LANGUAGE:** (English) *AND* **DOCUMENT TYPES:** (Article OR Review) *Indexes=SCI-EXPANDED, SSCI, A&HCI TIMESPAN=2007-2019*
#2	68	(TS=(carbon label*) AND TS=(service)) *AND* **LANGUAGE:** (English) *AND* **DOCUMENT TYPES:** (Article OR Review) *Indexes=SCI-EXPANDED, SSCI, A&HCI TIMESPAN=2007-2019*
#1	1601	(TS=(carbon label*) AND TS=(product)) *AND* **LANGUAGE:** (English) *AND* **DOCUMENT TYPES:** (Article OR Review) *Indexes=SCI-EXPANDED, SSCI, A&HCI TIMESPAN=2007-2019*

3. Results

3.1. Number of Publications

As shown in Figure 1, there were 175 publications in the period of 2007–2009 with a gradual increase. Particularly, there was a peak in 2016. Less than five papers published annually during the period of 2007 to 2009, which highlighted the infancy of carbon labeling scheme, as it was firstly issued by UK in 2006. From 2009 to 2011, the number of publications increased rapidly. During 2011–2013, it had remained unchanged as 10 publications annually. From 2014, the growth was significant towards a peak in 2016, which had an increase of 15.5 times comparing with the publications in 2007. A possible reason might be that the Paris Climate Agreement drives structural transformation of the global carbon markets, which calls for effective market-based policy tools, e.g. carbon labeling schemes, eco-labeling schemes, carbon trade mechanisms, to promote emissions reduction and energy transformation [20–22]. Since 2017, the number of publications had decreased to 20 and remained at a slow growth until 2019.

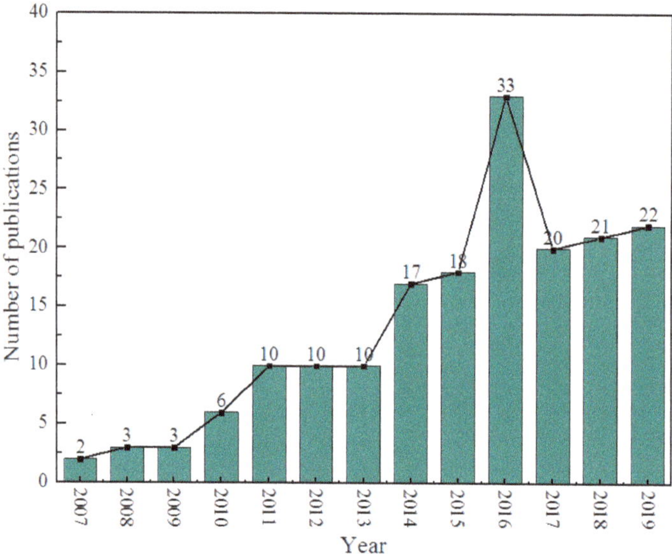

Figure 1. Output of publications during 2007–2019.

3.2. Countries of Publications

A total of 43 countries have contributed articles in carbon labeling related research areas. Table 2 shows the top 10 countries in terms of their publications during 2007–2019. It is apparent that U.S. ranks first, China as second, followed by UK, Germany, Australia, and Italy. In particular, European Union (EU) countries contributed most of the publications, indicating a regional spillover effect due to their close collaborations in economics and politics.

Table 2. Top 10 countries with the most published articles from 2007 to 2019.

Country	Centrality	TP	Percentage (%)
USA	0.98	41	23.43
China	0.21	27	15.43
UK	0.38	23	13.14
Germany	0.19	16	9.14
Australia	0.19	14	8.00
Italy	0.92	13	7.43
Netherland	0.70	10	5.71
Sweden	0.08	10	5.71
France	0.61	9	5.14
South Korea	0.82	7	4.00

TP: total publications.

Figure 2 shows the variation of publications regarding the top five countries, where there is a peak occurred between 2015 and 2018. United Kingdom and United States are the pioneers of carbon labeling schemes and have been involved in such studies for 12 years, indicating a remarkable spillover effect in policy design on low-carbon consumption and production. Carbon label related studies were still in progress in developing entities, e.g., China engaged in such issues until 2012. However, China makes great contribution in low-carbon development due to great pressure on its energy structure [23]. In such context, China calls for such a labeling system to change the consumption patterns while upgrading the supply chain to further improve product/service quality [24].

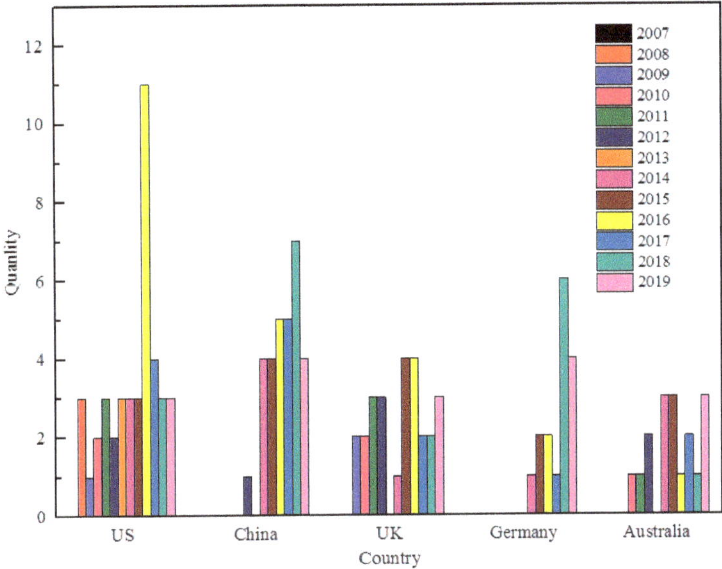

Figure 2. Number of publications regard the top five countries.

The national collaboration provides a new perspective to assess the academic impact of countries in carbon label related studies. Figure 3 shows the network of national collaboration, in which each country is denoted by a circle. The size of the circle represents the frequency of collaboration. The higher the frequency, the larger the circle. The thickness of the purple circle represents the centrality. The higher the centrality, the thicker the purple circle. The centrality is used to indicate the international status of a country in the carbon labeled studies. The US has the largest degree of centrality (0.98), implying a comparative high international influence in this research field. Italy, in spite of only 13 articles being published, its centrality ranks the second (0.92), indicating great potential in development of international cooperation. For instance, there were at least two authors from other countries in these 13 published papers. China, whilst ranked as the second in the number of publications, its centrality was relatively low, which further required reinforcement of international cooperation to share the research findings.

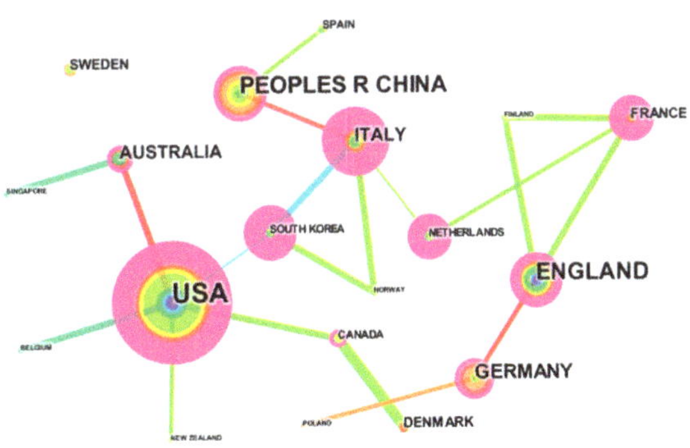

Figure 3. National cooperation network.

3.3. Subject Categories and Published Journals

There are wide topics involved in the studies of carbon labeling schemes during the period 2007–2019, which are subject to 44 categories. Figure 4 shows the top 10 categories, and the most common category is "Environmental Sciences", accounting for 23.1%. The subject categories are mainly related to environment, economy, food, and business, which demonstrate that carbon labeling related studies have a multidisciplinary field.

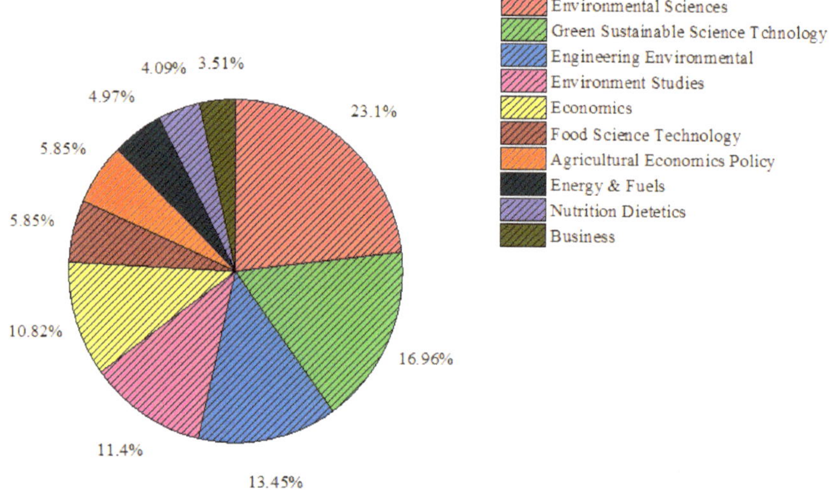

Figure 4. Top 10 subject categories.

Carbon label related studies were published in 85 journals. Table 3 shows the top 10 journals, where 86 articles are published, accounting for 49.14%. Most of the journals were from the publisher Elsevier. The *Journal of Cleaner Production* is highlighted by both of the maximum publications and the total citations. The *Journal of Renewable and Sustainable Energy Reviews* has the highest impact factor, and the *Journal of Food Policy* has the highest average number of citations.

Table 3. Top 10 journals with the most published articles.

Journals	IF(2018)	EP	Percentage (%)	TC	ACP
Journal of Cleaner Production	6.395	37	21.14	149	4.03
Food Policy	3.788	10	5.71	91	9.10
Sustainability	2.592	10	5.71	1	0.10
Energy Policy	4.880	6	3.43	3	0.50
International Journal of Life Cycle Assessment	4.868	5	2.86	7	1.40
British Food Journal	1.717	4	2.29	24	6.00
Appetite	3.501	4	2.29	19	4.75
Renewable & Sustainable Energy Reviews	10.556	4	2.29	14	3.50
Environmental Science & Policy	4.816	3	1.71	26	8.67
Energy Economics	4.151	3	1.71	21	7.00

EP: entire publications, TC: total citations, ACP: average citation per paper.

3.4. Distribution of Institutions

Table 4 shows the top five institutions ranked by their affiliated first author's publications. There were 287 institutions involved in the carbon label related studies during 2007–2019. Southwest Jiaotong University was identified as the institution where most of the publications contributed, followed by Central Queensland University, Swedish University of Agricultural Sciences, Eidgenössische Technische Hochschule Zürich (ETH), and Korea University. Southwest Jiaotong University published six papers in the area, where it mainly focused on application of system dynamics and game theory to investigate the interaction among consumer, enterprises, and government in implementation of the carbon labeling scheme to provide insightful policy implications on sustainable consumption and production [25,26]. Central Queensland University proposed carbon labeling scheme as an indicator to reflect the environmental impact of building materials in order to promote green design [27,28]. Swedish University of Agricultural Sciences was ranked as the fourth, where they paid close attention to measurement of uncertainties in carbon labelled food product [29,30]. ETH and Korea University gave emphasis on consumers' preferences and willingness to pay for carbon labeled products, by which the possible influencing factors were explored [31–33].

Table 4. Institutions with the most publications.

Institution	Country	TP
Southwest Jiaotong University	China	6
Central Queensland University	Australia	4
Swedish University of Agricultural Sciences	Sweden	4
Eidgenössische Technische Hochschule Zürich	Switzerland	4
Korea University	Korea	3

TP: total publications.

3.5. *Highly Cited Papers*

The top 10 highly cited papers in the field of carbon labeling schemes during the period 2007–2019 are given in Table 5. These papers were distributed in seven journals, and three of the highly cited papers were published on *Journal of Food Policy*. The article published by Grunert et al. [34] was ranked the first with 267 citations. In these highly cited papers, some focused on comparing utilities of various sustainable labeling schemes, including carbon, environmental, and ethical labels [31,35]. Most of the citations related to these highly cited papers were mainly from methodological perspectives, focusing on public attitudes towards organic foods, by means of choice experiments, structured interviews, and questionnaire survey [36,37]. The co-citations were highlighted by exploration of consumer preferences and willingness to pay for organic foods by using Schwartz's value theory combined with the planned behavior theory [38,39]. However, several studies firmly believed that carbon labels may provide a signal to help consumers change their purchasing behaviors if they can fully understand the associated labeling information [40,41]. In such context, Rugani et al. [42] argued if the underlying method for carbon labeling, i.e., the life cycle assessment, could be improved to be more transparency. In summary, these highly cited papers had addressed the challenges in application of carbon labeling schemes, to lay out a foundation for their future development.

3.6. *Keywords*

Keywords can be used to reflect the hotspots and topics of research interest in a certain time period [18]. CiteSpace software package was employed to produce a keywords co-occurrence network [43]. In such a process, a number of synonymous keywords were sorted by merging, such as "Carbon label" and "Carbon labeling", "greenhouse gas" and "GHG", "life cycle assessment" and "LCA", etc. Table 6 shows the frequency regarding the keywords occurred during 2007 to 2019. There were 98 keywords obtained, among which 22 keywords appeared above 10 times. During the time period of retrieval, "carbon footprint",' "willingness to pay",' and "food", were the top three keywords, indicating that surveys on consumer attitudes towards carbon labeled products had aroused widely academic concerns in the past 12 years.

Table 5. Top 10 cited papers.

Title	Author	Country and Institution	Journal and Year	TC	ACP
Sustainability labels on food products: Consumer motivation, understanding and use	Grunert, K.G.; Hieke, S; Wills, J.	Denmark, Aarhus University	*Food Policy* (2014)	267	44.50
Does local labeling complement or compete with other sustainable labels? A conjoint analysis of direct and joint values for fresh produce claim	Onozaka, Y.; McFadden, D.T.	Norway, University Stavanger	*American Journal of Agricultural Economics* (2011)	144	16.00
the use and usefulness of carbon labeling food: A policy perspective from a survey of UK supermarket shoppers	Gadema, Z.; Oglethorpe, D.	England, Northumbria University	*Food Policy* (2011)	112	12.44
Carbon labeling of grocery products: public perceptions and potential emissions reductions	Upham, P.; Dendler, L.; Bleda, M.	England, University Manchester	*Journal of Cleaner Production* (2011)	108	12.00
Consumers' valuation of sustainability labels on meat	Van Loo, E.J.; Caputo, V.; Nayga, R.M.; Verbeke, W.	South Korea, Korea University	*Food Policy* (2014)	85	14.17
Product-level carbon auditing of supply chains Environmental imperative or wasteful distraction?	McKinnon, A.C.	Scotland, Heriot Watt University	*International Journal of Physical Distribution & Logistics Management* (2010)	68	6.80
The potential role of carbon labeling in a green economy	Cohen, M.A.; Vandenbergh, M.P.	USA, Vanderbilt University	*Energy Economics* (2012)	67	8.38
Finnish consumer perceptions of carbon footprints and carbon labeling of food products	Hartikainen, H.; Roininen, T.; Katajajuuri, K.M.;Pulkkinen, H.;	Finland, MTT Agrifood Research Finland	*Journal of Cleaner Production* (2014)	64	10.67
Vulnerability of exporting nations to the development of a carbon label in the United Kingdom	Edwards-Jones, G.; Plassmann, K.; York, E.H.; Hounsome, B.; Jones, D.L.; Canals, L.	Wales, Bangor University	*Environmental Science & Policy* (2009)	64	5.82
Challenges of carbon labeling of food products: a consumer research perspective	Roos, E.; Tjarnemo, H.;	Sweden, Swedish University of Agricultural Sciences	*British Food Journal* (2011)	51	5.67

TC: total citations, ACP: average citation per paper.

Table 6. Descriptive statistics of the keywords.

Keywords	Count	Centrality	Percentage (%)
Carbon footprint	50	0.18	7.72
Willingness to pay	37	0.18	5.38
Food	28	0.03	4.07
Carbon label	26	0.38	3.78
Consumption	26	0.20	3.78
Product	24	0.07	3.49
Climate change	21	0.10	3.05
Information	20	0.32	2.91
Behavior	18	0.13	2.62
Greenhouse gas emission	18	0.04	2.62
Life cycle assessment	17	0.19	2.47
Eco label	16	0.21	2.33
Sustainability	16	0.23	2.33
Impact	15	0.06	2.18
Choice	14	0.12	2.03
Footprint	14	0.12	2.03
Perception	14	0.05	2.03
Attitude	13	0.20	1.89
Label	13	0.03	1.89
Choice experiment	12	0.08	1.74
Preference	11	0.10	1.60
Policy	10	0.14	1.45

Figure 5 shows the keywords co-occurrence network. The size of the circle represents the occurrence frequency. The lines between the nodes denote their connections. The thicker the lines, the stronger the connection. The lines between nodes are bright in color, indicating that there are a number of research hotspots derived in recent years. The largest circle is "carbon footprint", by which "carbon label," "willingness to pay," "food," and "attitude" are closed linked. The node with the highest centrality is "carbon label", by which there are six nodes connected, including "carbon footprint," "information," "life cycle assessment," "performance," etc. Such phenomenon may imply that life cycle-based assessment is fundamental to the performance of carbon labeling schemes, through which various forms of carbon footprint information is provided. Besides, consumer behavior towards a carbon labeling scheme is full of academic research interests. Consumers are receptors of carbon labeled products or services, and their purchase intentions are critical to implementation of the labeling policy [11,44].

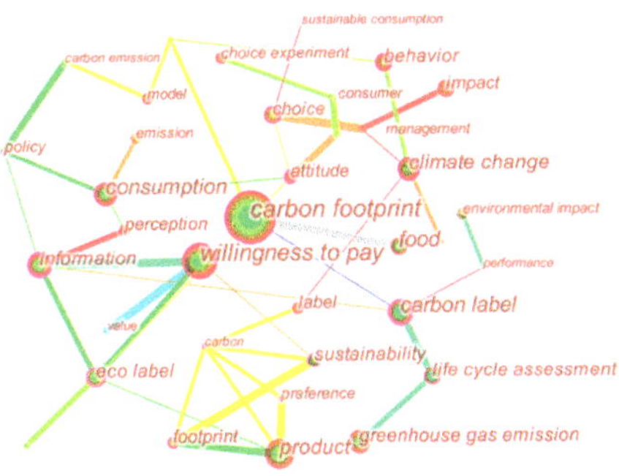

Figure 5. Keywords co-occurrence network.

Keywords clustering was conducted to identify the research frontiers of carbon label related studies. There were seven clusters identified, and their corresponding silhouette values were above 0.7 (if above 0.5, the cluster is considered to be reasonable), as shown in Figure 6. The largest cluster (#0) was entitled "categorisation task", which put emphasis on application of classification to ensure reliable comparisons among similar products with different carbon emissions. The second cluster (#1), was given as "uncertainty analysis", focusing on uncertainty regarding carbon footprint assessment. The third cluster (#2), as "carbon footprint label", mainly paid attention to its relative performance, e.g., the environmental impact, energy efficiency etc. The cluster (#3), as "climate change", indicating public awareness of the implication of carbon labeling scheme.

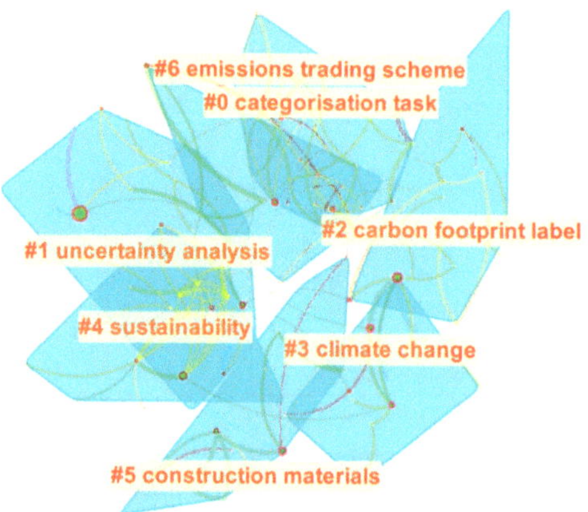

Figure 6. Clusters of carbon label studies related keywords.

Burst detection was used to determine if any change occurred at the research hotspots [45]. The bursting words refer to the keywords that have suddenly emerged or increased significantly in a short period of time, which may provide insight into identification of future research interests [46]. It is detected by an algorithm proposed by Kleinberg (2002), which generates a list of important words in terms of their frequencies in a finite duration of time [45]. The frequency of the word changes implies possible state transition, indicating as burst [47].

In Figure 7, the red rectangle indicates the strongest bursts, since the corresponding keywords have multiple occurrences in the specific time nodes. "Strength" shown in Figure 7, indicates the bursting words that have been mentioned frequently than any other words in a specific time period [19]. There are 15 keywords with apparent bursts in this study. Such phenomenon implies that the carbon labeling studies have been distinguished by three stages: first from 2007 to 2012, as the carbon labeling scheme was in its infancy, and the burst keywords mainly contained carbon footprint, carbon label, carbon emission, food, and energy. In particular, the labeling policy was gradually extended to the household equipment and construction industry to evaluate its energy efficiency. The second stage was from 2013 to 2015, where the keywords were booming, with life cycle assessment, eco label, food consumption, and market contained. In this stage, studies preferred the utility of the labeling policy and its possible impact on production, trade, and export. For instance, the carbon labeling schemes were compared with other eco-labels to highlight the impact on the development of trade and economy [48]. In the third stage (2016–2019), studies placed emphasis on individuals' behavior towards carbon labeled product or service by conducting surveys to investigate their perception, preference, and willingness to pay, indicating that they were interests of topics in the carbon label studies [49–52]. At the same time, multi-stakeholders' interaction was gradually involved in the carbon label relevant researches [53]. The application of a carbon labeling scheme was gradually transformed from product to service, e.g., tourism. Carbon labeling may affect the behavior of tourists who purchases low-carbon tourism services, though tourists may have pro-environmental intent or not, as well as may display environmentally sustainable behavior or not [54].

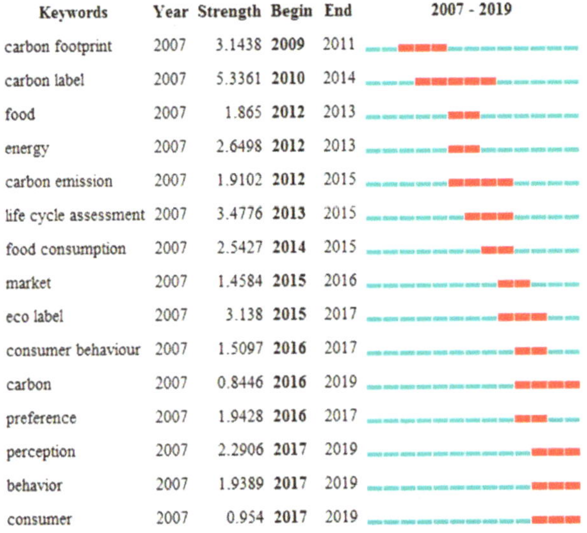

Keywords	Year	Strength	Begin	End	2007 - 2019
carbon footprint	2007	3.1438	**2009**	2011	
carbon label	2007	5.3361	**2010**	2014	
food	2007	1.865	**2012**	2013	
energy	2007	2.6498	**2012**	2013	
carbon emission	2007	1.9102	**2012**	2015	
life cycle assessment	2007	3.4776	**2013**	2015	
food consumption	2007	2.5427	**2014**	2015	
market	2007	1.4584	**2015**	2016	
eco label	2007	3.138	**2015**	2017	
consumer behaviour	2007	1.5097	**2016**	2017	
carbon	2007	0.8446	**2016**	2019	
preference	2007	1.9428	**2016**	2017	
perception	2007	2.2906	**2017**	2019	
behavior	2007	1.9389	**2017**	2019	
consumer	2007	0.954	**2017**	2019	

Figure 7. Top 15 Keywords ranked by burst detection.

4. Discussion and Insights

4.1. Discussion

This study highlights the necessity to make clearer the carbon labeling information and introduces a multiple-perspective co-citation analysis for interpreting co-citation cluster dynamics. Moreover, unlike other studies, the research visualizes a geographical as well as a multidisciplinary importance of carbon labeling schemes. The study underlines the novelty of a co-occurrence network among environments, economies, food, and businesses analyzed in different countries within the environmental science scenario. The research turns to improving network visualization to help consumers better understand the utility of labeling scheme.

The development of research hotspots of carbon labeling schemes has been characterized by the alteration of keywords during the time period 2007–2019. The alterations of keywords show that research aims were changing from carbon labeling as a marketing tool to consumption behavior and attitudes helpful for environmental policies. In recent years, academic circles have focused on consumption behavior in order to understand carbon labelling impact.

The study manages a bibliometric analysis with CiteSpace, which visualizes the evolution trend of carbon labeling scheme. The paper presents an upgrading of the pre-existing literature review, which stopped in 2016 [23]. From 2016 to 2019, scientific research paid more attention to consumer behavior, preference, and attitude towards carbon labeling. The keywords co-occurrence network highlights a linkage between life cycle assessment and carbon labeling, mainly for food and energy, showing academia's interest about willingness to pay, life cycle assessment, and carbon labeling. Furthermore, there is a strong connection between consumer attitude and consumption choice as well as sustainability label and eco-label. The burst detection highlighted that the investigation of consumer behavior was a topic of interest in the carbon labeling related research field.

Unlike previous studies, the paper takes into consideration countries, such as the U.S., China, and Italy. The literature on carbon labeling schemes coming from China highlights a renewed interest in this issue. China focuses on application of system dynamics investigating the interaction among consumers, enterprises, and governments in implementation of carbon labeling [25,26]. In the future, China might improve its carbon labeling policy and put it into wide practice, influencing market prospect, and providing insightful policy implications on sustainable consumption and production.

4.2. Insights

Though the analytic review on the 175 carbon label related articles, three main challenges have been identified to lay out foundations for future studies. First, precise accounting for carbon emissions is a prerequisite for the labeling practice [29]. Carbon label is also an entitled carbon footprint label, since most of the carbon labels are presented in a footprint form [6]. As implied in the definition of carbon label, lifecycle-based carbon footprinting is a cornerstone to support the presentation of a carbon labeling scheme [55]. However, the system boundary for a specific product or service is difficult to define, which may cause uncertainty in expression of the accounting results; thus, decreasing credibility of the labeling scheme [28]. For instance, existing studies have shown that the carbon footprint of crops are varied by place of production, due to the different system boundary for lifecycle accounting, giving rise to uncertainty in food carbon labeling practice [30]. Such inconsistency may give rise to the same product that has significantly different numerical values labeled on its package [56]. It, thus, calls for improvement of carbon footprint assessment to ensure fair comparison among similar products or services. Besides, a functional unit is generally followed by the life cycle assessment, which limits comparability among various types of products [29]. There is a call for standardized methods to normalize the carbon footprint (regarding products or services) into a common scale, in order to improve the comparability of various labeling schemes [57].

The second issue with the labeling practice is the poor communication with consumers. A number of studies have identified that consumers are confused by the labeling information, even though they are willing to pay certain premium for carbon labeled product or service [58,59]. In such context, research turns to improving visualization to help consumers better understand the utility of labeling scheme. For example, a traffic light colored system was proposed to indicate intensity of product carbon footprint by using the normalization method [60,61]. Whether such form of label suggested is effective in enhancing communication still needs further validation. Moreover, it is worth noting that consumers may be irrational regarding environmental concerns [62]. Conventional research methods, including questionnaires, focus groups, in-depth interviews, etc., have been widely employed to explore consumer perception and willingness to pay for carbon labeled products [63]. However, they may be limited by capturing responses based on consciousness [64,65]. Neuroscience is insightful to identify the conscious and subconscious responses; thus, to discriminate social consciousness and actual behaviors [66,67]. Such a tool has potentials to investigate consumer behavior towards different forms of labeling presentation.

The third issue is the labeling policy overlapping. Taking food as an example, there are a number of labels presented on its package, such as information regarding organic, food miles, animal welfare, and carbon footprint [68]. Various labels not only add complexity in the packaging design, but also give rise to the issues with respect to information credibility and reliability, even resulting in more confusion when consumers purchase products. Thus, the integration of various labeling policies is essential to ensure information coverage and improve labeling form of presentation [48,69]

5. Conclusion and Implications

This study reviews carbon label related studies in the period of 2007–2019 based on a bibliometric analysis. The number of publications, the countries of publications, the categories, the journals, the authors, the institutions, and the highly cited papers are investigated to have a holistic view on the research progress regarding carbon labeling scheme. There are 175 publications identified in the defined period, which presents an increase trend. The publications cover 44 categories and 85 journals. "Environmental Sciences" is the key subject category, and "*Journal of Cleaner Production*" is the journal with the most publications. 287 institutions have contributed to this research area. U.S. ranks the first in the number of publications, followed by China, UK, Germany, Australia, and Italy.

The study investigates the research progress of carbon labeling schemes, to provide insight into its future development. The outcomes confirm the necessity to optimize life-cycle assessment for carbon labeling schemes as well as the importance of labeling for better visualization. Moreover, the study gives support to the importance of improving carbon labeling schemes to help consumers change their purchasing behaviors.

Some considerations about the implications of this study are introduced. The first regards the cooperation between developing countries and developed countries. Spillover effects may accelerate carbon labeling scheme implementation. Developed countries may transfer experiences, good practices to developing countries to ameliorate carbon-labeling schemes, so as to standardize the schemes internationally. Carbon labels, in particular, and eco-labels, in general, represent a private governance and may enhance compliance. The standardization of the carbon labeling scheme at an international level may be useful to define a limit for the number of categories, as well as for the variety of labels that sometimes overlap each other [6].

The second implication takes into consideration the consumers and their motivation. It is necessary to define strategies to motivate consumers towards decreasing GHG emissions, and facilitate the knowledge of the labeling meanings. At this purpose, green education programs should be defined, involving both consumers and entrepreneurs, as well as eco-advertising, which, using, for instance, social media, encourage people to purchase environmentally friendly products and services.

The third implication concerns the use of new business models and technologies. The use of eco-branding, such as names and symbols, to distinguish environmentally friendly businesses from environmentally unfriendly competitors. The adoption of carbon-branding implements sustainable business practices leading to a reduction of costs, enhancing corporate social responsibility, attracting new markets, and obtaining higher prices [11]. Technologies, e.g., blockchain, which is a public digitalized ledger, may be used in the carbon labeling process to create more transparency and trust in the relationship between customers and sellers (concerning environmentally friendly purchasing).

There is a connection between carbon labeling schemes and life cycle assessment. Future research, thus, would be the optimization of life cycle assessment for labeling accreditation, improvement of labeling visualization for better expression, and normalization of various environmental labels to promote sustainable consumption. How to implement carbon label standards, as well as homogenizing carbon labeling schemes among different countries should be investigated to provide insightful policy implications on sustainable consumption and production. Moreover, future studies may observe that carbon labeling schemes trigger consumers to take pro-environmental purchasing actions. In such cases, the research might deeply analyze the linkages between consumers and enterprises, as well as between enterprises and governments, to implement carbon labeling schemes.

Author Contributions: R.Z. was involved in conceptualization of this study; D.W. implemented the bibliometric analysis; R.Z. prepared the manuscript draft; S.P. revised and edited the manuscript. All authors have read and agreed to the published version of the manuscript.

Funding: This study is sponsored by the National Natural Science Foundation of China (No. 41301639, 41571520), Sichuan Provincial Young Talent Program (No. 2019JDJQ0020), Chengdu Soft Science Fund (2020-RK00-00246-ZF), Sichuan Province Circular Economy Research Center Fund (No. XHJJ-2002).

Conflicts of Interest: The authors declare no conflict of interest.

References

1. Zhao, R.; Min, N.; Geng, Y.; He, Y.L. Allocation of carbon emissions among industries/sectors: An emission intensity reduction constrained approach. *J. Clean. Prod.* **2017**, *142*, 3083–3094. [CrossRef]
2. Natural Resources Defense Council. Available online: https://www.nrdc.org/resources/paris-agreement-climate-change (accessed on 17 July 2020).

3. Tan, M.Q.B.; Tan, R.B.H.; Khoo, H.H. Prospects of carbon labelling—A life cycle point of view. *J. Clean. Prod.* **2014**, *72*, 76–88. [CrossRef]
4. Carbon Trust. Available online: https://www.carbontrust.com/resources/the-carbon-emissions-generated-in-all-that-we-consume (accessed on 17 July 2020).
5. Bolwig, S.; Gibbon, P. Counting carbon in the marketplace: Part 1—Overview paper. In Proceedings of the Global Forum on Trade—Trade and Climate Change, Paris, France, 9–10 June 2009.
6. Liu, T.T.; Wang, Q.W.; Su, B. A review of carbon labeling: Standards, implementation, and impact. *Renew. Sustain. Energy Rev.* **2016**, *53*, 68–79. [CrossRef]
7. Mostafa, M.M. Egyptian consumers' willingness to pay for carbon-labeled products: A contingent valuation analysis of socio-economic factors. *J. Clean. Prod.* **2016**, *135*, 821–828. [CrossRef]
8. Ciasullo, M.V.; Maione, G.; Torre, C.; Troisi, O. What about sustainability? An empirical analysis of consumers' purchasing behavior in fashion context. *Sustainability* **2017**, *9*, 1617. [CrossRef]
9. Carbon Trust. Available online: https://www.carbontrust.com/resources/product-carbon-footprint-labelling-consumer-research-2020 (accessed on 17 July 2020).
10. Feucht, Y.; Zander, K. Consumers' preferences for carbon labels and the underlying reasoning. A mixed methods approach in 6 European countries. *J. Clean. Prod.* **2018**, *178*, 740–748. [CrossRef]
11. Shuai, C.; Ding, L.; Zhang, Y.; Guo, Q.; Shuai, J. How consumers are willing to pay for low-carbon products?—Results from a carbon-labeling scenario experiment in China. *J. Clean. Prod.* **2014**, *83*, 366–373. [CrossRef]
12. Li, Q.W.; Long, R.Y.; Chen, H. Empirical study of the willingness of consumers to purchase low-carbon products by considering carbon labels: A case study. *J. Clean. Prod.* **2017**, *161*, 1237–1250. [CrossRef]
13. Plassmann, K. Comparing voluntary sustainability initiatives and product carbon footprinting in the food sector, with a particular focus on environmental impacts and developing countries. *Dev. Policy Rev.* **2018**, *36*, 503–523. [CrossRef]
14. Zhao, R.; Liu, Y.Y.; Zhang, N.; Huang, T. An optimization model for green supply chain management by using a big data analytic. *J. Clean. Prod.* **2017**, *142*, 1085–1097. [CrossRef]
15. Gadema, Z.; Oglethorpe, D. The use and usefulness of carbon labelling food: A policy perspective from a survey of UK supermarket shoppers. *Food Policy* **2011**, *36*, 815–822. [CrossRef]
16. McKinnon, A.C. Product-level carbon auditing of supply chains. *Int. J. Phys. Distrib. Logist. Manag.* **2010**, *40*, 42–60. [CrossRef]
17. Shi, X.P. The spillover effects of carbon footprint labelling on less developed countries: The example of the East Asia Summit region. *Dev. Policy Rev.* **2013**, *31*, 239–254. [CrossRef]
18. Yang, Y.; Meng, G.F. A bibliometric analysis of comparative research on the evolution of international and Chinese ecological footprint research hotspots and frontiers since 2000. *Ecol. Indic.* **2019**, *102*, 650–665. [CrossRef]
19. Chen, C.M. Science mapping: A systematic review of the literature. *J. Data Inf. Sci.* **2017**, *2*, 1–40. [CrossRef]
20. Falkner, R. The Paris Agreement and the new logic of international climate politics. *Int. Aff.* **2016**, *92*, 1107–1125. [CrossRef]
21. Aldy, J.; Pizer, W.; Tavoni, M.; Reis, L.A.; Akimoto, K.; Blanford, G.; Carraro, C.; Clarke, L.E.; Edmonds, J.; Iyer, G.C.; et al. Economic tools to promote transparency and comparability in the Paris Agreement. *Nat. Clim. Chang.* **2016**, *6*, 1000–1004. [CrossRef]
22. Fujimori, S.; Kubota, I.; Dai, H.; Takahashi, K.; Hasegawa, T.; Liu, J.; Hijioka, Y.; Masui, T.; Takimi, M. Will international emissions trading help achieve the objectives of the Paris Agreement? *Environ. Res. Lett.* **2016**, *11*, 104001. [CrossRef]
23. Liu, W.L.; Wang, C.; Xie, X.; Mol, A.P.J.; Chen, J.N. Transition to a low-carbon city: Lessons learned from Suzhou in China. *Front. Environ. Sci. Eng.* **2012**, *6*, 373–386. [CrossRef]
24. Wang, C.X.; Wang, L.H.; Liu, X.L.; Du, C.; Ding, D.; Jia, J.; Yan, Y.; Wu, G. Carbon footprint of textile throughout its life cycle: A case study of Chinese cotton shirts. *J. Clean. Prod.* **2015**, *108*, 464–475. [CrossRef]
25. Zhao, R.; Han, J.J.; Zhong, S.Z.; Huang, Y. Interaction between enterprises and consumers in a market of carbon-labeled products: A game theoretical analysis. *Environ. Sci. Pollut. Res.* **2018**, *25*, 1394–1404. [CrossRef] [PubMed]
26. Zhao, R.; Zhou, X.; Han, J.; Liu, C. For the sustainable performance of the carbon reduction labeling policies under an evolutionary game simulation. *Technol. Forecast. Soc. Chang.* **2016**, *112*, 262–274. [CrossRef]

27. Wu, P.; Feng, Y. Using lean practices to improve current carbon labelling schemes for construction materials—A general framework. *J. Green Build.* **2012**, *7*, 173–191. [CrossRef]
28. Wu, P.; Low, S.P.; Xia, B.; Zuo, J. Achieving transparency in carbon labelling for construction materials—Lessons from current assessment standards and carbon labels. *Environ. Sci. Policy* **2014**, *44*, 11–25. [CrossRef]
29. Röös, E.; Sundberg, C.; Hansson, P. Uncertainties in the carbon footprint of food products: A case study on table potatoes. *Int. J. Life Cycle Assess.* **2010**, *15*, 478–488. [CrossRef]
30. Röös, E.; Sundberg, C.; Hansson, P. Uncertainties in the carbon footprint of refined wheat products: A case study on Swedish pasta. *Int. J. Life Cycle Assess.* **2011**, *16*, 338–350. [CrossRef]
31. Van Loo, E.J.; Caputo, V.; Nayga, R.M.; Verbeke, W. Consumers' valuation of sustainability labels on meat. *Food Policy* **2014**, *49*, 137–150. [CrossRef]
32. Lazzarini, G.A.; Visschers, V.H.M.; Siegrist, M. Our own country is best: Factors influencing consumers' sustainability perceptions of plant-based foods. *Food Qual. Prefer.* **2017**, *60*, 165–177. [CrossRef]
33. Shi, J.; Visschers, V.H.M.; Bumann, N.; Siegrist, M. Consumers' climate-impact estimations of different food products. *J. Clean. Prod.* **2018**, *172*, 1646–1653. [CrossRef]
34. Grunert, K.G.; Hieke, S.; Wills, J. Sustainability labels on food products: Consumer motivation, understanding and use. *Food Policy* **2014**, *44*, 177–189. [CrossRef]
35. Onozaka, Y.; McFadden, D.T. Does Local Labeling Complement or Compete with Other Sustainable Labels? A Conjoint Analysis of Direct and Joint Values for Fresh Produce Claim. *Am. J. Agric. Econ.* **2011**, *93*, 693–706. [CrossRef]
36. Vermeir, I.; Verbeke, W. Sustainable food consumption: Exploring the consumer "Attitude—Behavioral Intention" gap. *J. Argic. Environ. Ethics* **2006**, *19*, 169–194. [CrossRef]
37. Janssen, M.; Hamm, U. Product labelling in the market for organic food: Consumer preferences and willingness-to-pay for different organic certification logos. *Food Qual. Prefer.* **2012**, *25*, 9–22. [CrossRef]
38. Aertsens, J.; Verbeke, W.; Mondelaers, K.; Van Huylenbroeck, G. Personal determinants of organic food consumption: A review. *Br. Food J.* **2009**, *111*, 1140–1167. [CrossRef]
39. Zander, K.; Hamm, U. Consumer preferences for additional ethical attributes of organic food. *Food Qual. Prefer.* **2010**, *21*, 495–503. [CrossRef]
40. Peschel, A.O.; Grebitus, C.; Steiner, B.; Veeman, M. How does consumer knowledge affect environmentally sustainable choices? Evidence from a cross-country latent class analysis of food labels. *Appetite* **2016**, *106*, 78–91. [CrossRef]
41. Camilleri, A.R.; Larrick, R.P.; Hossain, S.; Patino-Echeverri, D. Consumers underestimate the emissions associated with food but are aided by labels. *Nat. Clim. Chang.* **2019**, *9*, 53–58. [CrossRef]
42. Rugani, B.; Vázquez-Rowe, I.; Benedetto, G.; Benetto, E. A comprehensive review of carbon footprint analysis as an extended environmental indicator in the wine sector. *J. Clean. Prod.* **2013**, *54*, 61–77. [CrossRef]
43. Chen, C.M.; Song, M. Visualizing a field of research: A methodology of systematic scientometric reviews. *PLoS ONE* **2019**, *14*, e0223994. [CrossRef]
44. Zhao, R.; Geng, Y.; Liu, Y.; Tao, X.; Xue, B. Consumers' perception, purchase intention, and willingness to pay for carbon-labeled products: A case study of Chengdu in China. *J. Clean. Prod.* **2018**, *171*, 1664–1671. [CrossRef]
45. Chen, C.M. CiteSpace II: Detecting and visualizing emerging trends and transient patterns in scientific literature. *J. Am. Soc. Inf. Sci. Technol.* **2006**, *57*, 359–377. [CrossRef]
46. Zhu, J.; Hua, W. Visualizing the knowledge domain of sustainable development research between 1987 and 2015: A bibliometric analysis. *Scientometrics* **2017**, *110*, 893–914. [CrossRef]
47. Kleinberg, J. Bursty and hierarchical structure in streams. *Data Min. Knowl. Disc.* **2003**, *7*, 373–397. [CrossRef]
48. Onozaka, Y.; Hu, W.; Thilmany, D.D. Can eco-labels reduce carbon emissions? Market-wide analysis of carbon labeling and locally grown fresh apples. *Agric. Food Syst.* **2016**, *31*, 122–138. [CrossRef]
49. Röös, E.; Tjarnemo, H. Challenges of carbon labelling of food products: A consumer research perspective. *Br. Food J.* **2011**, *113*, 982–996. [CrossRef]
50. Spaargaren, G.; Van Koppen, C.S.A.K.; Janssen, A.M.; Hendriksen, A.; Kolfschoten, C.J. Consumer responses to the carbon labelling of food: A real life experiment in a canteen practice. *Sociol. Rural.* **2013**, *53*, 432–453. [CrossRef]
51. Zhao, R.; Zhong, S.Z. Carbon labelling influences on consumers' behaviour: A system dynamics approach. *Ecol. Indic.* **2015**, *51*, 98–106. [CrossRef]

52. Chen, N.; Zhang, Z.H.; Huang, S.M.; Zheng, L. Chinese consumer responses to carbon labeling: Evidence from experimental auctions. *J. Environ. Plan. Manag.* **2018**, *61*, 2319–2337. [CrossRef]
53. Zhao, R.; Zhou, X.; Jin, Q.; Wang, Y.; Liu, C.L. Enterprises' compliance with government carbon reduction labelling policy using a system dynamics approach. *J. Clean. Prod.* **2017**, *163*, 303–319. [CrossRef]
54. Juvan, E.; Dolnicar, S. Measuring environmentally sustainable tourist behaviour. *Ann. Tour. Res.* **2016**, *59*, 30–44. [CrossRef]
55. Hu, A.H.; Chen, C.H.; Lan, Y.C.; Hong, M.Y.; Kuo, C.H. Carbon-labeling implementation in Taiwan by combining strength-weakness-opportunity-threat and analytic network processes. *Environ. Eng. Sci.* **2019**, *36*, 541–550. [CrossRef]
56. Cohen, M.A.; Vandenbergh, M.P. The potential role of carbon labeling in a green economy. *Energy Econ.* **2012**, *34*, S53–S63. [CrossRef]
57. Zhao, R.; Deutz, P.; Neighbour, G.; McGuire, M. Carbon emissions intensity ratio: An indicator for an improved carbon labelling scheme. *Environ. Res. Lett.* **2012**, *7*, 014014. [CrossRef]
58. Upham, P.; Dendler, L.; Bleda, M. Carbon labelling of grocery products: Public perceptions and potential emissions reductions. *J. Clean. Prod.* **2011**, *19*, 348–355. [CrossRef]
59. Koistinen, L.; Pouta, E.; Heikkilä, J.; Forsman-Hugg, S.; Kotro, J.; Mäkelä, J.; Niva, M. The impact of fat content, production methods and carbon footprint information on consumer preferences for minced meat. *Food Qual. Prefer.* **2013**, *29*, 126–136. [CrossRef]
60. Osman, M.; Thornton, K. Traffic light labelling of meals to promote sustainable consumption and healthy eating. *Appetite* **2019**, *138*, 60–71. [CrossRef]
61. Thøgersen, J.; Nielsen, K.S. A better carbon footprint label. *J. Clean. Prod.* **2016**, *125*, 86–94. [CrossRef]
62. Hornibrook, S.; May, C.; Fearne, A. Sustainable Development and the Consumer: Exploring the role of carbon labelling in retail supply chains. *Bus. Strategy Environ.* **2015**, *24*, 266–276. [CrossRef]
63. Lombardi, G.V.; Berni, R.; Rocchi, B. Environmental friendly food. Choice experiment to assess consumer's attitude toward "climate neutral" milk: The role of communication. *J. Clean. Prod.* **2017**, *142*, 257–262. [CrossRef]
64. Matukin, M.; Ohme, R.; Boshoff, C. Toward a better understanding of advertising stimuli processing exploring the link between consumers' eye fixation and their subconscious responses. *J. Advert. Res.* **2016**, *56*, 205–216. [CrossRef]
65. Van Gaal, S.; Lamme, V.A.; Fahrenfort, J.J.; Ridderinkhof, K.R. Dissociable brain mechanisms underlying the conscious and unconscious control of behavior. *J. Cogn. Neurosci.* **2011**, *23*, 91–105. [CrossRef] [PubMed]
66. Khushaba, R.N.; Wise, C.; Kodagoda, S.; Louviere, J.; Kahn, B.E.; Townsend, C. Consumer neuroscience: Assessing the brain response to marketing stimuli using electroencephalogram (EEG) and eye tracking. *Expert Syst. Appl.* **2013**, *40*, 3803–3812. [CrossRef]
67. Zhao, R. Neuroscience as an insightful decision support tool for sustainable development. *Iran. J. Public Health* **2019**, *48*, 1933–1934. [CrossRef] [PubMed]
68. De-Magistris, T.; Gracia, A.; Barreiro-Hurle, J. Do consumers care about European food labels? An empirical evaluation using best-worst method. *Br. Food J.* **2017**, *119*, 2698–2711. [CrossRef]
69. Galli, A.; Wiedmann, T.; Ercin, E.; Knoblauch, D.; Ewing, B.; Giljum, S. Integrating ecological, carbon and water footprint into a "footprint family" of indicators: Definition and role in tracking human pressure on the planet. *Ecol. Indic.* **2012**, *16*, 100–112. [CrossRef]

© 2020 by the authors. Licensee MDPI, Basel, Switzerland. This article is an open access article distributed under the terms and conditions of the Creative Commons Attribution (CC BY) license (http://creativecommons.org/licenses/by/4.0/).

MDPI
St. Alban-Anlage 66
4052 Basel
Switzerland
Tel. +41 61 683 77 34
Fax +41 61 302 89 18
www.mdpi.com

Energies Editorial Office
E-mail: energies@mdpi.com
www.mdpi.com/journal/energies

www.ingramcontent.com/pod-product-compliance
Lightning Source LLC
LaVergne TN
LVHW070631170726
843515LV00004B/229